Speed of light
186,000 /miles per second

Light year = 5.6 trillion miles
electric magnetic spectrum

What is light connection to

✗ waves x Ray - Gamma
   occilatins

   red to violet
micro waves, radio waves
to x Rays . Gammas

W9-CCR-004

EDGE OF THE UNIVERSE

A Voyage to the Cosmic Horizon and Beyond

Paul Halpern, PhD

WILEY

John Wiley & Sons, Inc.

Dedicated, with love,
to Felicia, Eli, and Aden

Copyright © 2012 by Paul Halpern. All rights reserved

Cover Image: © iStockphoto
Cover Design: Jose Alamaguer

Published by John Wiley & Sons, Inc., Hoboken, New Jersey

No part of this publication may be reproduced, stored in a retrieval system, or transmitted in any form or by any means, electronic, mechanical, photocopying, recording, scanning, or otherwise, except as permitted under Section 107 or 108 of the 1976 United States Copyright Act, without either the prior written permission of the Publisher, or authorization through payment of the appropriate per-copy fee to the Copyright Clearance Center, 222 Rosewood Drive, Danvers, MA 01923, (978) 750–8400, fax (978) 646–8600, or on the web at www.copyright.com. Requests to the Publisher for permission should be addressed to the Permissions Department, John Wiley & Sons, Inc., 111 River Street, Hoboken, NJ 07030, (201) 748–6011, fax (201) 748–6008, or online at http://www.wiley.com/go/permissions.

Limit of Liability/Disclaimer of Warranty: While the publisher and the author have used their best efforts in preparing this book, they make no representations or warranties with respect to the accuracy or completeness of the contents of this book and specifically disclaim any implied warranties of merchantability or fitness for a particular purpose. No warranty may be created or extended by sales representatives or written sales materials. The advice and strategies contained herein may not be suitable for your situation. You should consult with a professional where appropriate. Neither the publisher nor the author shall be liable for any loss of profit or any other commercial damages, including but not limited to special, incidental, consequential, or other damages.

ISBN 978-0-470-63624-4

Printed in the United States of America

Contents

Prologue

Cosmology's Extraordinary New Frontiers

Everything that we once believed about the universe is wrong! We thought that most of the material in space was made up of atoms, or at least of visible substances. Wrong! We thought that the expansion of the universe was slowing down—that its growth from the Big Bang was losing steam. Wrong again! We thought that galaxies were fairly evenly spread out and that there were no large regions of space without them. That was before the discovery of giant voids rendered that notion null and . . . wrong once more! We thought that there was a single universe—after all, "uni" means one. While the jury is still out on that one, some scientists are already claiming evidence of a multiverse—a collection of parallel realities. So ultimately even the name "universe" could turn out to be wrong!

Welcome to twenty-first-century cosmology, a highly precise field that is unafraid to admit that the vast majority of the universe is made of things beyond our current understanding. Dark matter, dark energy, and vast voids in space are the much-discussed cosmological topics of today that, like the 1990s television series *Seinfeld,* are all "about nothing." Nothingness is literally on cosmologists' radar screens, as they are learning about the enormous gaps through comprehensive sky surveys by space-based and ground telescopes as well as by detailed probes of the radio signals left over from the dawn of time, among other sources.

Clearly these things aren't really "nothing"—we just don't know enough to say what they are. Traditionally, astronomers have focused on what can be directly observed—the stuff of stars and planets. They estimate that about 4 to 5 percent of the universe is composed of conventional matter. But for those following the latest trends, ordinary matter is *so* twentieth-century. We've been there, done that, and now wish to tackle the exquisite enigma of fathoming the unseen majority of the cosmos.

As the science of the entire physical universe, cosmology shifts its goals and scope in tandem with the ebb and flow of knowledge about space. Through modern techniques, many great mysteries that once baffled philosophers and scientists have been resolved. All manner of cosmological data point to a primordial era of the cosmos that was incredibly hot and unbelievably dense, called the Big Bang. Astronomers have homed in on the approximate age of the universe, 13.75 billion years (give or take 100 million years or so), and have developed a detailed picture of some of its early stages. Recent models estimate the size of the observable universe (the part potentially detectable by instruments that measure signals from space) to be approximately 93 billion light-years in diameter. Yet, ironically, it seems that as we learn more and more about the cosmos, we realize how much of it lies beyond our grasp.

Five centuries ago intrepid European explorers set sail beyond the sea's visible horizon and mapped out lands hitherto unknown to them. Today, astronomers have embarked upon an even more extraordinary quest: to gauge the shape, horizons, and extent of the universe itself, including the enormous invisible regions. For this cosmic voyage, compasses, sextants, and parchment scrolls have been replaced by powerful telescopes, delicate microwave receivers, sophisticated computer algorithms, and a host of other tools for capturing light waves from across the spectrum. The emerging cartography of the universe is being pieced together from its immensely intricate record of luminous signals—even as we attempt to penetrate the secrets of the darkness. In this sunrise of the new cosmology, it is the multifarious hues of collected, scientifically analyzed light that will illuminate the endless night.

Darkness surrounds our world, broken only by scattered points of light. We are used to the void, and undaunted by the task of gleaning information from far-flung objects that only dimly announce their arrival each night. Our adeptness with fashioning light-gathering mirrors and lenses has served us well, enabling us to map out parts of space from which signals literally take billions of years to reach us. Astronomy is now comfortable with such challenges.

Yet we have grown increasingly cognizant of a far more perplexing darkness that so far has defied all attempts at understanding. While sheer nothingness without effect is easy to dismiss, invisible material that exerts unseen influence cannot be taken lightly. More and more, we realize that the things we observe are guided by substances we cannot see. The Milky Way and other galaxies have at their cores supermassive black holes, are steered by dark matter, and are driven apart by dark energy. These are three different types of cloaked entities. The latter two provide, when added up, the lion's share of all the matter and energy in the universe.

At least we know what conventional black holes are made of—highly compressed material states formed when the cores of massive stars catastrophically collapse. They are so gravitationally intense that nothing can escape their grasp—not even light signals. Supermassive black holes—much larger than the ordinary variety—formed from remnants of earlier, more massive stellar generations and likely played major roles in how galaxies such as the Milky Way assembled themselves. Thus, although black holes are mysterious, astronomers have constructed viable models about how they emerged.

We cannot say the same for dark matter; no one knows its actual composition. Astronomers infer the existence of dark matter because of the behavior of stars in galaxies and of galaxies in clusters. Invisible material tugs on stars in the outer reaches of galaxies, forcing them to revolve around the galactic centers much faster than they would otherwise. Furthermore, without the gravitational "glue" from unseen substances, galaxies would not be able to form the giant clusters we see in the sky, such as the Coma Cluster, the Virgo Cluster, and numerous others. Recent studies of dark matter around clusters have shown that it is typically shaped like an elongated cigar rather than a perfectly symmetric sphere. Scientists estimate that approximately 23 percent of all the stuff in the universe is made of dark matter. Yet, despite multiple experiments, its identity remains unknown.

An even greater chunk of everything in the observed universe is made of a completely different yet equally mysterious entity called dark energy. Unlike dark matter's unseen "glue," dark energy exerts an invisible repulsion—pushing galaxies away from one another. We've known since the late 1920s that the universe is expanding; however, it was only in the late 1990s that astronomers discovered that this cosmic expansion has been speeding up rather than slowing down. Nobody knows the cause of this acceleration. Possibly it is a modification of the law of gravity itself rather than an actual substance. Astrophysicist Michael Turner coined the term dark energy to distinguish it from dark matter, yet to reflect the elusive nature of both types of substances. Astronomers have estimated that more than 72 percent of everything in the universe is comprised of dark energy. The hefty fractions allotted to dark matter and dark energy mean that less than 5 percent of the universe is made of ordinary matter—the stuff of atoms and everything we see. Resolving the nature of dark matter and dark energy constitutes two of the greatest scientific conundrums of the modern age.

Limits to cosmic knowledge frustrate our spirit of curiosity. We wish to fathom all of physical reality and, like the readers of an incomplete novel, are baffled by the glaring omissions. Many other questions about the universe defy current attempts at understanding and challenge us to press further until we know the answers. Was there a beginning to time? Were there events before the Big Bang? Is it possible to travel backward into the past? Are there other universes? Could there be higher dimensions beyond our perception? Will the history of the cosmos ever draw to a close? If so, what will happen in its final era?

In recent years the idea of a multiverse—a collection of universes— has gathered much currency. A model of rapid growth in the very early universe, called inflation, suggests in some of its versions that our universe is surrounded by other, "bubble" universes. While the concept sounds like science fiction, amazingly, a recent discovery seems to bolster the idea of universes beyond ours.

In 2008 astronomer Alexander Kashlinsky of NASA's Goddard Space Flight Center announced the astounding results of a detailed study of the motions of clusters of galaxies. Using data gathered by the WMAP (Wilkinson Microwave Anisotropy Probe) satellite, his team found that hundreds of clusters are streaming through space at millions of miles per hour toward a patch of sky between the constellations Centaurus and Vela. The researchers speculated that this "dark flow," as they called this phenomenon, could be the result of material from beyond the observable universe affecting clusters within it through mutual gravitational attraction. Unseen parts of the universe, it seems, could be tugging on our coat sleeves, trying to let us know they are out there.

The statistical analysis of the microwave background radiation collected by WMAP and other instruments has proven to be a revolutionary tool for astronomy. It has revealed many oddities, including a lining up of patterns of ripples in such a strange way that it has been dubbed the "axis of evil." Another source of befuddlement has been the finding of several large cold spots. These are sizable patches of the sky, comparable in width to the full moon, where the temperature of the microwave radiation is lower, on average, than in other sectors. The significance of these cold spots is unclear. While some scientists dismiss them as mere statistical flukes, others have speculated that they could represent scars from interactions with other universes.

If that isn't weird enough, another cosmic mystery has come to the fore in the opposite end of the radiation spectrum. While WMAP has been mapping out subtle temperature differences of low-energy microwaves, another space instrument, the Fermi Gamma-Ray Space Telescope, has been surveying gamma radiation, the most energetic form of light. Whenever a massive star catastrophically bursts in what is called a supernova explosion, it releases colossal amounts of energy—more than what the Sun has released so far in its lifetime. Much of this energy is in the form of gamma rays. These gamma-ray bursts punctuate a background haze of such radiation, known as the gamma-ray fog. Until recently, astronomers have assumed that the gamma-ray fog is simply the sum total of bursts from distant galaxies—including the incessant output of supermassive black holes at their galactic centers, churning out floods of gamma rays as they gobble up matter. However, results from the Fermi Telescope released in early 2010 indicated that known sources made up only 30 percent of the gamma-ray fog. Fully 70 percent could not be explained—an oddity that the report called "dragons." What kinds of astral beasts could be hiding in the fog, breathing gamma-ray fire? Along with dark matter, dark energy, and dark flow, it is another great cosmic mystery.

Prepare yourself for an epic voyage to the cosmic horizon and beyond! Keep a watchful eye out for strange creatures lurking in the mist. Brace yourself for forces powerful enough to forge entire universes and for energies corrosive enough to obliterate them. And sail with utmost care around the black holes, lest you meet a plunging fate.

In our journey into the heart of space we have inherited the brave legacy of the mariners Leif Erikson and Magellan, of the great Polynesian explorers who sailed the Pacific with longboats, of the bold voyagers who crossed the Bering land bridge into North America, and of all those others seeking new lands and adventure. At present we travel with our eyes, our instruments, and our imagination rather than our physical bodies; but who knows about the far future?

This is a tale of cosmic dragons, bottomless pits, and looking-glass worlds—of a possible axis of evil and hypothesized portals to hidden realms. Darkness pervades this story—dark matter, dark energy, and dark flow—but it is in essence a chronicle of how radiant light is collected from the far reaches of space, broken up into its kaleidoscope of colors and invisible frequencies, and analyzed by probing minds. From these images and theories has emerged a narrative even more exciting than the heroic sagas of yore. Let the celestial adventure begin!

1

How Far Out Can
We See?

Voyage to the Edge of the
Known Universe

High in heaven it shone,
Alive with all the thoughts, and hopes, and dreams
Of man's adventurous mind.
Up there, I knew
The explorers of the sky, the pioneers
Of science, now made ready to attack
That darkness once again and win new worlds.

—ALFRED NOYES, "WATCHERS OF THE SKY" (1922)

Modern science suggests that space is infinite. Astronomical measurements have revealed that its geometry is flat, like an endless plane, but in three dimensions. It's truly a mind-bending concept, because if the universe is infinitely large, we are infinitely small.

However, when scientists describe the contents of the universe—including its stars, galaxies, and other features—they are speaking of the knowable universe, the observable universe. What lies beyond that edge we can only assume. We cannot peek beyond the borders of the observable universe—no matter how good our telescopes and other measuring instruments get—so we don't know what fraction of the actual physical universe it represents. While we might speculate that space extends forever, we cannot prove so definitively. So how large is this observable universe, our enclave of darkness and light that we will never see beyond? Scientists have estimated it to be about 93 billion light-years across. Their ability to make such a far-reaching statement points to stunning advances in astronomical measurement.

Seeing Back in Time

Cosmology, the science of the universe, is witnessing a golden age. Thanks to powerful telescopes, delicate microwave receivers, sophisticated computer algorithms, and a host of other tools for collecting and analyzing light, the study of the cosmos has achieved extraordinary precision. Researchers have been able to extend the reach of astronomical knowledge out to unprecedented depths and farther into the past than ever before. Ultimately, our ability to see the edge of the knowable universe is bounded by how far back in time we can observe.

From this remarkable body of collected data we can trace the growth of the universe backward in time and pin down its age. Some 13.75 billion years ago, everything we see around us, from Earth to the outer limits of observation, burst forth in the fiery, ultracompact event known as the Big Bang. To be able to address with such authority something that took place so long ago speaks to modern cosmology's triumph of precision.

Let's estimate the size of the observable universe. As a first approximation, for simplicity's sake, let's pretend that the universe has been static since its beginning. Imagine that at its birth it simply popped into being the way it is right now. The distance from here to the edge of the observable universe would be its age multiplied by the speed of light. Light races through a vacuum at about 186,000 miles per second—somewhat less than 6 trillion miles per year. For convenience astronomers refer to the distance light travels in a year as a light-year, about 5.9 trillion miles. When we look at an object 1 light-year away, we are actually seeing it as it was a year ago: it took one year for light to cross that light-year. When we look at an object 100 light-years away, we are seeing it as it was before you were born. When we look at the edge of the known universe, we are not seeing it as it is now, but as it was in its nascent era. Therefore, in the static scenario, the farthest object we could see would be approximately 13.75 billion light-years (about 81 billion trillion miles) away, because that would be the distance light could possibly travel since the beginning of time. That would be the extent of the knowable universe.

However, because the universe is expanding, we are able to see bodies that are much farther out in space. That is because after an object emits light, space continues to grow—propelling the object farther away from us. By the time we receive the signal, expansion has rendered the luminous body much more remote. Moreover, the expansion of space has been speeding up, pushing the body even farther away. Consequently, the size of the observable universe is much larger than it would have been if space had kept still.

It was only in 2005 that noted Princeton astrophysicist J. Richard Gott III and his colleagues calculated the radius of the observable universe to be approximately 46.6 billion light-years.[1] Its diameter would be double that, or about 93 billion light-years across. No instrument, no matter how powerful, could detect a signal from farther away.

Compared to the hypothetical scenario of perfect instruments, however, the reach of telescopes is somewhat more limited. Telescopes collect light that has traveled through space after being emitted by various sources. According to Gott and his collaborators, the radius of the sphere containing all the potentially detectable light sources is approximately 45.7 billion light-years. They performed this calculation by extrapolating back in time to the recombination era, determining the limits of the farthest light sources from that period that we could presently observe, and using the expansion rate of the universe and other cosmological data to determine where those sources are today.

The recombination era is the age, some 380,000 years after the Big Bang, when atoms formed and light began to travel freely through space. Light from that era is the earliest we can detect. Before then, the universe was opaque, meaning that light was unable to travel far. Light bounced back and forth among electrons, positive ions (atoms missing electrons), and other charged particles. Thus, the difference between the two values of the radius lies in whether or not it will ever be possible to record non-luminous signals from the opaque period between the Big Bang and the recombination era. If we include that possibility, the observable universe is about 46.6 billion light-years in radius. If we consider only light sources, the figure becomes 45.7 billion light-years in radius.

It is strange to imagine the passage of time during the opaque period, given that light did not move much through space. Therefore, there is no visual record of what happened during those 380,000 years or so. Without motion it's hard to picture time. However, the clock ticking during that interval was the growth of space itself. By measuring the material composition of the universe and relying on Einstein's general theory of relativity to determine how the expansion rate has changed over time, contemporary scientists have been able to determine how long it took from the Big Bang until the recombination era, when the universe was cool enough for atoms to form.

Only after the electrons and ions formed neutral atoms did the universe become transparent to light. Light could stop bouncing from particle to particle and start moving in a straight line. The liberated photons (light particles) could journey freely through the cosmos, traveling at the speed of light over great distances, offering a more familiar way to measure time by dividing distance by velocity. Some of that light has reached us today in the form of cosmic microwave background radiation—a radio hiss that permeates space.

Because the universe has become larger and larger since its beginning, Gott and his coworkers took into account how its scale has changed over time. For the expansion rate and other cosmological parameters, they used pinpoint results from NASA's Wilkinson Microwave Anisotropy Probe (WMAP) satellite, designed to gather information about the cosmic microwave background. Analysis of the WMAP data has yielded clear-cut values for long-sought quantities, such as the age, shape, expansion rate, and acceleration of the universe. Until that information was available, astronomical estimates of the size of the observable universe had been far less precise. Thanks to the data, the researchers were able to calculate a much firmer value for the knowable universe's extent.

Gott and his colleagues estimated in their paper that the observable universe is home to some 170 billion galaxies, housing, all in all, some 60 sextillion (billion trillion) stars. More recent estimates are even higher—maybe even as many as 1 trillion galaxies, encompassing, in total, hundreds of sextillions of stars. The immensity of observable space truly boggles the mind! It is amazing that modern telescopes have been able to probe a significant fraction of such a vast expanse.

Journey from the Center of the Universe

When we consider what is at the edge of the observable universe we might also wonder what is at its center. A common myth is that there is some place in space where the Big Bang took place. However, there is no such point because cosmic growth takes place equally everywhere in space. The expansion due to the Big Bang pushes all points in space away from their neighboring points. If we trace that growth backward in time, we find that all points get closer and closer together. Therefore, there is no single place in the universe that uniquely defines where the Big Bang occurred.

However, the observable universe does indeed have a center. How could we find it? Is there a marker at that point, like the Four Corners Monument, where Arizona, New Mexico, Colorado, and Utah meet? Is it a busy, well-trafficked hub, like Piccadilly Circus in London and the Colosseum in Rome? Or is it a remote place, hard to access, but of great significance because of the conjunction of lines sketched by cartographers, like the South Pole?

As a matter of fact, you could stand at any of those places and rightly shout, "I'm at the center of the observable universe!" By definition, any viewing point in space is the center of the observable universe because light arrives at that location equally from all directions. It is like being in a boat on the open sea and seeing horizon all around; wherever the boat is at the time would be the center of the circle traced by that horizon. So if you set up a telescope in Times Square, on top of Mount Everest, or at the peak of the highest mountain in Hawaii, Mauna Kea, each of those places would be the universe's center for you. (If choosing among these, Mauna Kea would be best for astronomy, as it would be far from city lights, high up, but still accessible by road. That's why there are important observatories right there.)

All this might seem a matter of semantics. We might wonder where the center of the actual physical universe is, not just the observable universe. In that case, scientists believe that there is no such point. The existence of such a center would defy a commonsense rule in astronomy: the Copernican Principle.

The Copernican Principle extrapolates the revolutionary idea of sixteenth-century astronomer Nicolaus Copernicus—that Earth is not the center of everything and that the Sun and planets do not revolve around it—to the universe as a whole. It offers a powerful tool for logically deducing the properties of other regions of the universe by assuming that our part of space is not special. For example, the distribution of galaxies we see should be about the same as an onlooker would observe at the same time from another planetary system in another galaxy 1 billion light-years away. A universe in which every locale looks pretty much the same is called homogeneous.

By making telescopic observations in all directions we might make an even stronger statement. On the largest scale, at every angle in which we gaze, what we see looks approximately the same in terms of the distribution of matter. That is called isotropic. If by sheer coincidence we were indeed at the actual center of the physical universe, we could attribute the observed isotropy to us being in a kind of hub, like at the center of a traffic circle in which streets radiate in all directions. However, if we follow the wisdom of Copernicus and presume that we aren't central, isotropic here would mean isotropic everywhere. The result is an extremely strong statement about the geometry of space, useful to our understanding of cosmology. The universe, on its largest scale, is homogeneous and isotropic.

Light's Extraordinary Tale

From Earth's humble vantage point, it is remarkable that we know any of this about the immense space around us. After all, manned space missions have ventured so far only to the Moon, a mere footstep away in the cosmic perspective (though a giant leap according to us!). Robotic probes have explored the solar system but haven't yet bridged the vast divide between the Sun's minions and the domain of the stars. Our physical journeys have touched only the tiniest sliver of the observable universe.

Fortunately, the incessant rain of light particles, or photons, down on Earth offers us a deluge of information about the cosmos. As telescopes have become increasingly sophisticated, equipped with colossal, precisely ground mirrors and high-resolution digital cameras, they have been better able to gather this flood of luminous data for subsequent analysis.

What can a photon tell us? For one thing, it moves in a straight line at a constant velocity—the speed of light—unless it encounters matter. Light can be emitted or absorbed by any type of substance than is electrically charged. This can alter its path and slow it down. For example, a negatively charged electron can give off a photon that is later picked up by another electron. This banter is part of the electromagnetic interaction, the process by which electric and magnetic forces are conveyed. That is why light is called an electromagnetic wave.

Another property of light is its brightness. The observed brightness of a luminous object, such as a streetlamp or star, depends on two main factors. One is the luminosity, or how much power the shining body puts out at its source. The other factor is the distance between the light source and the observer. The distance factor is an inverse-squared relationship, meaning that if you stand twice as far from a lightbulb, it appears four times dimmer. While it might take a centrally located 100-watt lightbulb to illuminate a room that is 30 feet across, enlarge the space to 300 feet across (the size of a football field) and it would take a 10,000-watt lightbulb at the same place to provided similar brightness. On the other hand, if you were sitting in an otherwise dark stadium and noticed the dim glow of a lamp that you knew was emitting substantial power, you'd realize that you were very far away from it. You might even use your knowledge of its luminosity, or actual power, and its apparent brightness to deduce how far away it was.

Astronomers use this comparison between luminosity and observed brightness through objects of known power output called standard candles. Standard candles offer an important way of gauging large distances. They have been used to establish the immensity of the universe and to help determine the rate by which it is expanding.

Among the oldest and most commonly used standard candles are stars called Cepheid variables. These blink regularly in brightness like a flickering lightbulb on a Christmas tree. In 1908 Henrietta Leavitt, working at Harvard College Observatory, made a monumental discovery about Cepheids that would change the course of astronomical history. She charted the blinking rate versus the luminosity of a group of Cepheids in the Magellanic Clouds (now known to be satellite galaxies of the Milky Way) and found a direct correlation. The more powerful the variable star, the slower it blinked. It was as if 100-watt lightbulbs, in flickering, had a longer cycle than 75-watt lightbulbs. Through meticulous studies, Leavitt established a clear connection between length of blinking cycle and luminosity for every Cepheid she observed. The Cepheid's calculated luminosity could then be compared to its observed brilliance to gauge how far away it was from Earth. The dimmer a Cepheid of known luminosity appeared, the farther away it was. Using this technique, Cepheids would prove faithful yardsticks for measuring astronomical distances.

To add another instrument to our scientific tool kit, let's consider yet another feature of light. Not only is it swift and bright, it also has colors. These varied hues relate to a property called frequency, or the number of cycles per second. Electromagnetic waves, as they travel through space, oscillate at different rates. The frequency of these waves can vary significantly—a palette of possibilities called the electromagnetic spectrum. The most familiar form of electromagnetic radiation, visible light, has a variety of frequencies ranging from approximately 400 trillion to 750 trillion Hertz (cycles per second). Our eyes perceive these frequencies as a rainbow of colors ranging from red on the low-frequency end to violet on the high-frequency end.

There's much more to light, however, than the eye can see. Of lower frequency than red is infrared radiation, which can be picked up though certain types of night-vision goggles and cameras. Even lower in frequency than that are microwaves and radio waves, used in communication technologies.

Moving to higher frequencies from violet is ultraviolet, familiar as the invisible cause of tanning and sunburn. Oscillating at an even faster rate are X-rays, used in many kinds of medical imaging. Capping off the electromagnetic spectrum are ultra-high-frequency gamma rays.

If you look at a shining object such as a lamp, it might seem to be only a single color or simply white. However, if you place in front of it a prism or a diffraction grating (two types of optical devices that break up light into its component frequencies), you are likely to see lines of various hues, each with a characteristic brightness. For some types of objects only certain colors might be represented—green and yellow, for instance, but not blue.

Astronomers rely on spectral lines to discern much information about the makeup of stars and other radiant objects. By analyzing light from astronomical bodies, researchers are able to discern their apparent percentage of hydrogen, helium, and other gases. Each element has its own characteristic spectral pattern, which readily identifies it. With the discoveries of planets in other systems, a new astronomical mission is to map out the composition of alien worlds as well as stars.

Light spectra have another important use besides identifying chemical composition. Through the Doppler effect, they also serve as reliable speedometers. The Doppler effect is the shortening of the measured wavelengths of light sources moving toward an observer and the lengthening of the wavelengths of sources moving away from an observer. Wavelengths are the distance between peaks of waves. They have an inverse relationship with frequency—the shorter the wavelength, the higher the frequency. When a light source is approaching, its waves get "squeezed," and its wavelength shortens. Its frequency shifts higher—toward the blue end of the spectrum. Scientists call this a Doppler blueshift. Conversely, when a source is moving farther away, its light waves stretch out, its wavelength lengthens, and its frequency shifts toward the lower, redder end of the spectrum—hence a Doppler redshift. By gauging the amount of blue or red shifting, observers can determine the speed of the light source either toward or away, respectively.

An analogy with sound waves helps illustrate this principle. Suppose a police car is racing toward a crime scene. Its siren wails at a higher pitch. After the criminal is apprehended and the car speeds away, its siren bellows lower. If you possessed ultrasensitive equipment for detecting the positions of spectral lines and aimed it at the car's headlights, you also might be able to detect a blueshift as the car is approaching and a redshift as it leaves the scene. Because light is far, far faster than sound, the effect would be much, much subtler—which is why our eyes wouldn't notice it.

If, on the other hand, the criminal were to speed away in his own stolen vehicle, the police could use a Doppler radar gun to measure how fast he was going. A radar gun bounces signals off of cars and other moving things, collects the returning waves, and measures their frequency difference from the outgoing waves. This comparison reveals the velocity of speeders and in some cases can be used as evidence against them.

Astronomers employ similar methods to assess how fast stars and other bodies are moving in comparison to Earth. One limitation of the Doppler method is that it determines speeds toward us or away from us but not along other directions. Therefore, if a star were traveling solely along a path perpendicular to astronomers' line of vision, they couldn't detect its speed using the Doppler technique.

In 1915, astronomer Vesto Slipher published a compendium of Doppler shift data of objects then called nebulae, including Andromeda and various other spiral-shaped and elliptical objects. He had gathered these data over the previous three years, starting with a 1912 spectroscopic observation of Andromeda. At that time, science had yet to establish the existence of other galaxies beyond the Milky Way. The astronomical community was unsure at the time whether these nebulae constituted interstellar gaseous clouds within the Milky Way or "island universes" in their own right. A heated debate raged about their distance, size, and significance. Soon American astronomer Edwin Hubble would reveal the answer.

Hubble and the Expanding Universe

Hubble was a man of many contradictions. Born in 1889 to a lower-middle-class family in rural Marshfield, Missouri, the "Top of the Ozarks," he developed into an ardent Anglophile after a Rhodes Scholarship offered him a stay at Oxford. He became a lifelong pipe-smoker and tweed jacket aficionado, and affected a posh English accent that struck some of his colleagues as pompous. Yet there was nothing pretentious about his research. Every detail he published was solid and careful—backed by meticulously collected astronomical data.

In 1919, the prominent American astronomer George Ellery Hale invited Hubble to helm the newly constructed Hooker Telescope at Mount Wilson Observatory. With its 100-inch mirror, it was at the time the largest telescope in the world. Nestled in the San Gabriel Mountains of Southern California, the center was noted for its clear, nighttime viewing sky at a time before nearby Los Angeles exuded a nocturnal electric glow. Hubble adeptly aimed the optical instrument at Cepheids in Andromeda, using these variable stars as standard candles to measure their distances from Earth. With these cosmic yardsticks, he demonstrated that Andromeda was far beyond the outer edge of the Milky Way. Based upon Andromeda's apparent diameter, he estimated its actual size and concluded that it was a galaxy in its own right. Like the Milky Way, it was full of stars. Soon it would be renamed the Andromeda Galaxy instead of its old designation, the Andromeda Nebula. Using the Cepheid method, Hubble charted the distances to other so-called nebulae and demonstrated that many of them were galaxies as well.

When Hubble published his findings in 1924, our concept of the size and content of the universe would be transformed forever. From that point forward we knew that the Milky Way is a mere droplet in the sea of reality and that our solar system is but a tiny speck in that droplet. Human significance in the scheme of everything shrank like a block of ice in a scalding vat. Humility would be our lot, thanks to the epic work of a proud astronomer.

After revealing that the universe extends far, far beyond the boundaries of the Milky Way, Hubble then took his Cepheid results, plotted them against Slipher's Doppler shift data, and noticed a startling pattern. Except for relatively close galactic neighbors such as Andromeda, all of the galaxies in space had redshifted light, with the amount of redshift increasing with distance. He plotted the velocities of these galaxies versus distance and drew a slanted line connecting the points. Hubble announced his discovery in 1929, and his conclusion was unmistakable—the farther away a galaxy, the faster it appeared to be fleeing from ours. Since it is ludicrous to think that our galaxy is some kind of pariah, he surmised that, except for nearest neighbors, all the galaxies in space are receding (moving away) from each other. We now call this the Hubble expansion, and the velocity-distance relationship, Hubble's law. The rate of galactic recession at any given time is called the Hubble constant—somewhat of a misnomer, as it can change from cosmic era to era.

Hatching the Cosmic Egg

Even before Hubble formulated his law of galactic recession, several theorists anticipated the idea that the universe is expanding. They drew their conjectures from Albert Einstein's masterful general theory of relativity, published in 1915. General relativity is a way of understanding gravity in which the matter and energy in a region of space distort the geometry of that region, causing it to warp, ripple, and bend into a variety of possible shapes. The more convoluted the mass distribution, the bumpier the fabric of that sector becomes. Consequently, any objects traveling through that region would be deflected by the lumps they encountered. As noted theorist John Wheeler (writing with his former students Charles Misner and Kip Thorne) succinctly described, "Space tells matter how to move. Matter tells space how to curve."[2]

To envision Einstein's theory, picture empty space as a soft mattress. Scatter small stones on its top and it would slightly indent under each. Drop a large rock on top and it would noticeably sag. A heavy enough boulder might even tear its fabric. Wherever there were dents, other objects moving along the top would travel along bent paths. A marble initially propelled in a straight line along a flat part of the mattress would nevertheless follow a curved trajectory once it reached the warped areas. That's because the shortest possible route in the distorted region would be curved rather than linear.

Similarly, general relativity mandates that objects moving through space and attempting to follow the shortest possible route would follow paths curved by the distribution of matter and energy. If you've ever wondered why the planets travel in elliptical orbits around the Sun, now you know: those are the straightest lines they can follow in the gravitational well created by the Sun's mass.

In 1922, Russian mathematician Alexander Friedmann solved Einstein's equations of general relativity for the universe as a whole. He chose the three geometries possible for an isotropic, homogeneous universe: flat, hyperspherical, and hyperboloid. Flat universes correspond to a standard Euclidean geometry in which parallel lines never meet and continue straight forever. It represents the three-dimensional equivalent of a plane. A hypersphere generalizes a sphere into a higher dimension. Just as a sphere has a two-dimensional exterior, such as the surface of Earth, a hypersphere has a three-dimensional exterior that curves around and connects up with itself. Similarly, a hyperboloid, or saddle shape, is the three-dimensional surface of the higher-dimensional equvalent of a hyperbola. Because of their forms, hyperspheres are said to be "closed" and hyperboloids are said to be "open."

Geometric laws offer clear distinctions among the three different possibilities. Flat universes would embrace the familiar rule that parallel lines maintain the same separation distance forever, never getting closer together or farther away. If the universe were shaped like a hypersphere, in contrast, parallel lines would always meet, like the lines of longitude on Earth converging at the North and South Poles. Finally, if the universe had a hyperboloid geometry, parallel lines would diverge like the folds of a clutched fan. A shorthand way of classifying these geometries is that a flat geometry has zero curvature; a closed, hypersphere geometry has positive curvature; and an open, hyperboloid geometry has negative curvature.

Friedmann plugged all three geometries into Einstein's equations of general relativity. The solutions Friedmann found indicated that in each case space would expand outward from a point. He did notice one important difference: while the flat and open geometries would expand forever, the closed geometry would grow for a time, stop expanding, reverse course, and begin contracting back down to a point. The third case would resemble a higher-dimensional version of a balloon growing as it is filled with air and then shrinking as the air is released.

These three scenarios bear a direct relationship to a parameter called omega. Omega is the ratio of the density of matter in the universe over a critical value that can readily be calculated. Omega can be less than, greater than, or equal to 1. The open case, with omega less than 1, corresponds to an underdense cosmos—one with insufficient matter ever to reverse course and contract down again. The closed situation, with omega greater than 1, matches up with an overdense cosmos. Under the weight of its own gravity, it would eventually cease expansion and start to collapse, leading to the crushing endgame that has come to be known as the "Big Crunch." Finally, the flat possibility, with omega equal to 1, would expand forever, albeit eternally approaching the brink of collapse—like an increasingly exhausted runner who still manages to soldier on. In today's evocative language, the open and flat final-stage scenarios have been deemed the "Big Whimper."

Visualizing the realistic dawn or demise of the actual physical universe, however, was the farthest thing from Friedmann's mind. He was interested in these scenarios as mathematical models rather than as representations of nature. Nevertheless, his work has proved extraordinarily important. His models have become standard portraits of the most uniform types of cosmic evolution derived from general relativity without a cosmological constant. Sadly, he died in 1925 at age thirty-seven, before his ideas became widely known.

A far more prominent advocate of an expanding universe was the Belgian mathematician and priest Georges Lemaître. Working independently, without any knowledge of Friedmann's conclusions, in 1927 Lemaître similarly solved Einstein's equations of general relativity and demonstrated how space could grow in size. After Hubble announced his results, Lemaître set out to refine the theory of an expanding universe and use it to develop a scientific story of creation. He imagined that the cosmos began as a "primeval atom"—a supercompressed state containing all matter—which he also called the "cosmic egg." This burst forth in a great explosion, released material that became the galaxies, and produced everything we see today. The recession of galaxies, Lemaître concluded, was the lasting legacy of the original blast.

Before Hubble's discovery of cosmic expansion, Einstein was unimpressed with expanding-universe scenarios. Rather, he believed in a stable universe. Although he had developed a dynamic theory of gravity, he had hoped that the overall effects of all the matter in the universe would somehow keep it from either growing or shrinking. He was flummoxed when that turned out not to be the case, and promptly added to his equations an ad hoc term called the cosmological constant, which would stabilize space and prevent it from expanding or contracting. The cosmological constant is like an emergency brake applied to the universe to keep it from rolling toward instability because its internal mechanisms, described by general relativity, failed to do so.

After Hubble announced his discovery, Einstein quickly reversed course. Einstein realized that his original model, without a cosmological constant, was more suitable for a growing universe. To reproduce the gallop of galaxies sprinting out of bounds, there was no need to rein in the equations with an extra stability term. He would call the addition of the cosmological constant his greatest blunder.

In 1931 Einstein visited Mount Wilson, met Hubble, and congratulated him on his achievement. It was a splendid meeting between the two figures who shaped modern cosmology, molding it into an exact science. Neither scientist had wanted a dynamic cosmos, yet the power of their methods had propped open its engine for the whole world to see. We've been in a great race since the dawn of time and didn't even know it until Hubble revealed the truth—which matched so well with Einstein's brilliant theory.

2

How Was the Universe Born?

Revealing the Dawn of Time

These observational results are in some ways so disturbing that
there is a natural hesitation in accepting them at their face value.
But they have not come upon us like a bolt from the blue, since
theorists for the last fifteen years have been half expecting that
a study of the most remote objects of the universe might yield a
rather sensational development.

—ARTHUR EDDINGTON, *THE EXPANDING UNIVERSE* (1933)

The universe has grown immensely since the time of the Big Bang. As
Princeton astrophysicist J. Richard Gott and his collaborators calculated, the observable universe—the part we can potentially detect—is
currently about 93 billion light-years in diameter. To perform this calculation they also determined how big this region was at the end of
the "recombination era," the moment some 380,000 years after the Big
Bang when electrically neutral atoms formed and radiation became free
to move through space. As we'll see, it was more than a thousand times

smaller in diameter, and more than a billion times smaller in volume, reaching just 85 million light-years across. Such tremendous growth is staggering to imagine.

The notion that space itself is expanding represents one of the most radical transformations in scientific thinking. From the time of the ancient Greeks until the early twentieth century, space was considered a fixed backdrop against which the cosmic drama played out. Newtonian physics, for example, suggested that while stars could move through space, its large-scale features would remain the same—a concept called absolute space. Hubble's observational findings, in contrast, paired with the theoretical work of Einstein, Friedmann, and Lemaître, pointed to a dynamic cosmos that grew from a point. A debate would develop between those who accepted the idea of a cosmic beginning and those who held onto the old idea that everything stays more or less the same over time. For the scientific community to unite behind the idea that the universe emerged from a superdense embryonic state billions of years ago would require more proof.

Evidence for the Big Bang, as it came to be called, would require more theoretical work related to how matter and energy developed in the hot, nascent cosmos. This would lead to a prediction that space today is filled with the relic radiation from the early universe cooled down over aeons to frigid temperatures only a few degrees above absolute zero. This cosmic background radiation would be mapped out during the latter decades of the twentieth century and the early part of the twenty-first, offering unmistakable proof of a fiery early stage of the universe and providing a bevy of information about its character.

The notion of how everything we see today emerged from simpler substances dates back to the 1930s, an age of great discovery in nuclear physics. Scientists began to realize that all the elements in the universe were built of similar ingredients—protons and neutrons at their core, and clouds of electrons surrounding these nuclei. Thanks to the pioneering work of Hans Bethe and others, based on an earlier suggestion of Eddington's, science learned how heavier elements could be built up from lighter ones in thermonuclear processes. Transformation of one element into another turned from an alchemist's dream into experimental reality. These transmutations could take place under conditions of extreme heat and intense pressure, such as the cores of stars, and explained the source of stellar energy. During the Second World War, the efforts of many nuclear scientists were diverted toward the military implications of atomic power. Afterward, attention turned once again to fundamental questions, including the role of nuclear physics in fueling astronomical processes.

There are two main types of energy-releasing nuclear transformations: fusion and fission. Fusion is the union of lighter elements and isotopes (variations of elements with different amounts of neutrons) into heavier ones. For example, under certain conditions, two nuclei of ordinary hydrogen can fuse into deuterium, also known as heavy hydrogen. While the cores of ordinary hydrogen are single protons, deuterium nuclei are proton-neutron pairs. In the process of binding together, one of the protons transforms into a neutron, releasing a positron (like an electron, but positively charged) and a neutrino (a lightweight neutral particle). Positrons are a form of antimatter, similar to matter but oppositely charged. They almost immediately run into electrons, mutually annihilate, and release two photons. Deuterium can fuse with protons to form a light form of helium called helium-3, which in turn can merge with more protons to create ordinary helium. The net result of the cycle is the transformation of hydrogen into helium, the release of neutrinos, and the production of energy. This is the main source of sunshine.

Fission occurs when some of the heaviest elements and isotopes, such as uranium-235, which has 92 protons and 143 neutrons, split into lighter ones. Like fusion, it also produces energy, as witnessed by residents who and industries that draw their power from commercial nuclear plants. The decay of radioactive isotopes is also the major source of heating Earth's core.

With fusion building up from the smallest elements and fission breaking down the largest, both types of nuclear processes tend toward the creation of moderate-size nuclei. The middle ground is iron, the most stable type of nucleus in nature. Transforming hydrogen into helium, and then (through another fusion process) into the next element, lithium, and so forth all the way up to iron, release energy. However, to transmute elements lower than iron into even higher elements, such as cobalt, nickel, copper, and zinc, requires the addition of energy. Even at 28 million degrees Fahrenheit (16 million degrees Celsius), the Sun's core isn't hot enough to do the trick. Where, then, are higher elements forged?

George Gamow, a pioneering Russian nuclear scientist born in Odessa, thought he knew the answer. Having attending university in St. Petersburg (then Leningrad), where he studied under Friedmann, he was familiar with cosmological models. He also had spent time in Copenhagen and Cambridge, with groundbreaking physicists such as Niels Bohr and Ernest Rutherford, and had developed an important model of a type of nuclear decay, before fleeing the Soviet Union and escaping to the West. Therefore he had the perfect background to contemplate relationships between the realms of the very tiny and the extremely large. In 1934 he became a professor of physics at George Washington University, where he continued his studies of nuclear processes. After the Second World War, his interests switched to cosmology and he began to ponder connections between element-building and the early universe. Working with PhD student Ralph Alpher, he proposed a remarkable theory of how everything in the universe came to be. The solution, they asserted, was that all the elements were forged in the cauldron of an extraordinarily hot and dense cosmic genesis.

Published in 1948, Alpher and Gamow's paper[1] had the unusual distinction of listing a third author who hadn't contributed at all. Gamow, who had a marvelous sense of humor, placed physicist Hans Bethe's name between Alpher's and his, to create a sequence sounding like the first three letters of the Greek alphabet: alpha, beta, and gamma. A further meaning is that he had hoped to explain how the chemical elements came about in succession, like the ordering of an alphabet.

Gamow designated the name "ylem" for what Lemaître had dubbed the "primeval atom." Along with Alpher and physicist Robert Herman, Gamow proposed that the ylem was extraordinarily hot and dense and was composed of freely moving protons, neutrons, and other particles. Because of the extremely high temperature, the particles were in an agitated state, cajoling one another like frenzied fans in a mosh pit. Randomly, protons and neutrons paired to form deuterium nuclei. Bombarded with additional neutrons, some of these couples transmuted into helium-3, then into ordinary helium, lithium, and so forth. In short order, according to the team's theory, all the natural elements formed. All the while, the universe was expanding. Thermodynamics teaches us that when things expand, they cool. Within minutes, the universe was cool enough that all the elements formed would no longer break apart—"frozen" into stability by the lack of thermal energy. These primordial components, according to Gamow and his colleagues, formed the seeds of all structures in the universe.

Today virtually no one refers to the idea of a fiery, compact nascent universe as "ylem," "primeval atom," or "cosmic egg." The name that stuck was "Big Bang." Curiously, that term was first used derisively. On a 1949 BBC Radio program, Cambridge astronomer Fred Hoyle dismissed the idea that all the material in the universe emerged from nothingness in a single "Big Bang." He considered the concept nonscientific because it violated the long-standing conservation law in physics that massive amounts of matter and energy cannot simply appear out of the blue. Debating the subject again and again, Hoyle would remain a leading critic of the Big Bang theory throughout his whole life.

Hoyle's alternative, developed along with astronomers Thomas Gold and Herman Bondi, was called the "Steady State" hypothesis and based on the principle of "continuous creation." It posited that as the galaxies in the universe move apart, new matter is slowly created to fill in the gaps. Eventually these form new galaxies, leaving the overall look of the cosmos essentially the same. Thus, although galaxies are receding, the universe was never very small.

Reportedly, the astronomical trio first discussed the notion of an endless universe in 1946 after seeing a horror movie called *Dead of Night*, which had a repeating nightmare scene. Gold wondered if a film could be made with a looping plot that made sense whenever you happened to start watching it. If you walked into the cinema an hour late, you could still begin watching the movie, stay an hour later, and miss none of the experience because there would be no real start or finish. Could the universe, the astronomers wondered, be just like that—having no true beginning or end?

To show how Steady State was a superior approach, Hoyle stole some of the thunder of the Big Bang theory by pointing out a major flaw in its story of element creation in a primordial fireball. He showed that there were missing rungs in the ladder; there were no stable isotopes with atomic weights of 5 or 8. Atomic weight is a measure of the number of protons plus the number of neutrons—together called nucleons.

Without stable combinations of 5 or 8 nucleons, there would not be an opportunity for higher elements to be formed before the universe cooled. It would be like trying to run up the stairs from the fourth to the sixth floor of a building if the fifth floor landing was completely blocked off and there was no time for an alternative route. Element-building would be stuck at lower levels. The cosmos would be left with hydrogen, helium, and a trace of lithium, but nothing else.

Gamow tried to develop an explanation for how element creation could leap the gap but was unsuccessful. On that account Hoyle turned out to be right. As he demonstrated in a 1957 paper[2] cowritten with astronomers Margaret and Geoffrey Burbidge and William Fowler, the production of higher elements actually takes place in the fiery, ultra-dense cores of giant stars. These are released into space when the giants catastrophically burst in what are called supernova explosions. Iron, carbon, and all the other chemical elements in our bodies—and in everything we see around us, except for hydrogen and a portion of the helium—were once inside stellar furnaces. Given that iron is an essential ingredient for hemoglobin, star stuff is literally in our blood.

The Big Bang did make much of the helium in the universe, however—a prediction that matches nicely with its known abundance. That matchup was a major success for the theory. Therefore, when helium balloons pop we might fondly recall the Big Bang, but when fireworks blast (or any other chemical processes occur), we should offer credit where credit is due and toast the supernovae that created the full gamut of chemical elements.

Blast from the Past

Throughout the 1950s and early 1960s, debate raged between the Big Bang and Steady State camps. While explaining element production was an important victory for Hoyle and company, it did not depend on or exclude a particular cosmological model. Another key prediction Gamow, Alpher, and Herman made using the Big Bang theory languished in obscurity until it was finally proven in the mid-1960s. Estimating the temperature of radiation released after the formation of atoms in the early universe, they used the expansion of the cosmos to calculate its rate of cooling. From that information they predicted that the universe currently should be full of relic radiation, cooled to only a frigid few degrees above absolute zero (their estimates of what this temperature would be were not very exact, as the age of the universe was then unknown).

In 1964, unaware of that prediction, two radio astronomers, Arno Penzias and Robert Wilson, were taking measurements with the giant Horn antenna at Bell Labs in Holmdel, New Jersey, when they made an extraordinary discovery. Originally built for radio communication, the Horn antenna was converted to astronomical use after the Telstar communications satellite was launched. Penzias and Wilson were searching with the antenna for signals from galaxies when they noticed a persistent hiss that seemed constant no matter which way it was pointed. After ruling out obvious candidates for interference, such as radio interference from nearby New York City, they checked it to see if pigeon droppings—which they referred to as "white dielectric material"—might be causing the problem. As it was hard to keep the birds away, the antenna was full of that stuff. However, cleaning out the antenna did nothing to mute the hum.

The way electromagnetic radiation makes itself known—from radio waves to visible light to gamma-rays—depends on the temperature of its source. The brightest rays of the hot Sun bathe us in yellow light (along with lesser quantities of other colors, ultraviolet, infrared, and so forth). Much, much colder things put out radio waves and microwaves. Penzias and Wilson mapped out the signals they detected, plotted the distribution—which centered in the microwave range—and estimated a temperature of approximately 3 degrees Kelvin (above absolute zero)—or minus 454 degrees Fahrenheit. What on Earth could be so cold? The answer lay billions of years before Earth was even created—in the frigid remnants of a once-fiery cosmic furnace.

Penzias and Wilson were like radio listeners who had stumbled upon the ultimate oldies station with the widest possible broadcast range. They had tuned in to the call letters B-A-N-G—airing signals billions of years old, round the clock, to all points in the cosmos. It was the ultimate blast from the past.

Robert Dicke of Princeton was the quintessential cosmologist. He was adept at developing clever theories and finding imaginative ways to test them. Before Penzias and Wilson announced their findings, he had just thought of a scheme to test the Big Bang hypothesis by measuring its relic radiation. Once Penzias informed him about the hiss he knew he had been beaten to a major find. Dicke ascertained that Penzias and Wilson had discovered the signature of the Big Bang etched in microwave background static. Dicke, along with Princeton astrophysicists P. J. E. (Jim) Peebles, Peter G. Roll, and David T. (Dave) Wilkinson, published a key paper announcing this result[3], followed in the same journal by a paper by Penzias and Wilson detailing their discovery.[4]

Dicke's scenario matched many of the predictions that Gamow, Alpher, and Herman had made years earlier. Dicke postulated that at a particular stage in the history of the universe, some 380,000 years after the Big Bang and well before the first stars formed, atomic nuclei and electrons assembled themselves into electrically neutral atoms. This represented a cosmic turning point—which, as we've discussed, has come to be known as the recombination era. While before that period, space was full of charged particles, protons and electrons, that volleyed photons between them, afterward light was much freer to move. The universe transformed from opaque to clear.

We can imagine the photons, during this transition, as acting a bit like the waiters at a crowded dinner party for singles (representing unpaired particles). During cocktail hour, with people wandering around the room looking for someone to talk with, the waiters might find it hard to get very far. Each step they took, they might bump into a partygoer and have to offer him or her hors d'oeuvres. Now imagine a recombination era where, fueled by delicious food and sparkling conservation (or at least a sparkling beverage), each single finds his or her perfect match. After all the partygoers pair up and sit down at tables to chat, the floor would become much freer for the waiters to move.

The radiation released during the recombination era was immensely hot—some 3,000 degrees Kelvin (close to 5,000 degrees Fahrenheit)—but it has cooled down to approximately 3 degrees K over the more than 13-billion-year interval between that time and the present. Because it once filled all of space, it continues to fill all of space, though space has greatly expanded.

According to the recent calculation by Gott and his collaborators that we discussed earlier, at the end of the recombination era the region that is now the observable universe was approximately 85 million light-years in diameter, about 1,090 times smaller than it is today.[5] That corresponds to a volume of the observable universe about 1.3 billion times smaller than at present. Given the enormous growth of space since that time, and the known effect that expansion produces cooling, it is no wonder that the radiation has cooled to such a frigid temperature.

Penzias and Wilson found the relic radiation to be smooth and even. No matter which way their antenna pointed, it had about the same temperature, within the limits of measurement. This overall uniformity proved that it was a feature of the entire universe, rather than of particular objects within it, such as galaxies. The discovery of the cosmic microwave background (CMB) radiation tipped the balance of the great cosmology debate—leading the bulk of the scientific community to embrace the Big Bang theory and abandon interest in the Steady State hypothesis as a credible alternative. While Hoyle and his colleagues continued to argue against the Big Bang, they became an increasingly isolated minority. In 1978, Penzias and Wilson were awarded a Nobel Prize in physics in honor of their outstanding achievement.

As the decades progressed after Penzias and Wilson's discovery, cosmologists began to raise a deeper question. Could the CMB radiation be too smooth? Where were the slight deviations in temperature expected for radiation produced during a phase of the universe that must have had some bumps?

This issue arose because nature is never absolutely uniform. You can see this by comparing a CGI (computer-generated) film to one with human actors. If computer-generated actors appear in a film, they often seem almost too perfect to be real. Close-ups typically don't reveal the subtle variations in skin tone that real actors would have. Similarly, the CMB didn't seem, on the face of it, to have any "blemishes." These would be expected from the way it was produced. During the recombination era, the radiation would have emanated from material constituents that could have been more dilute in some regions and denser in others. In fact, researchers argued, the primordial universe must have had such imperfections to form the seeds of the astronomical structures we see today, such as stars, galaxies, and clusters of galaxies. The variations in density in the early universe would have led to temperature differences in the radiation that eventually cooled to become the CMB.

In short, astronomers expected that the unevenness needed to form the star-speckled sky would similarly turn up in its relic radiation profile.

The link between primordial density variations and temperature fluctuations in the background radiation has to do with several different competing factors. First, radiation packed into a tighter space is hotter than if it is more spread out. The inside of an oven is hotter if the door is closed and the heated air is trapped inside than if the air is allowed to escape. Places where the matter particles were more clustered and the photons similarly bunched would therefore start with a higher temperature.

Another factor, called the Sachs-Wolfe effect, proposed by American astrophysicists Rainer Sachs and Arthur Wolfe in 1967, has to do with light particles escaping the gravitational "wells" dug out by the matter clumps. By "wells" we mean regions in which particles need energy to climb out. The deeper the well, the greater the amount of energy a particle must expend to free itself. Denser regions of matter have more concentrated gravitational influences and thus carve out deeper wells. It is like the footprints of elephants on wet sand. Just as an ant must burn more calories in climbing out of such an indentation than in, say, strolling away from a pigeon track, a photon must work harder to escape the gravity of more massive regions than those that are relatively sparse. The result is that the photons emanating from clumped areas would have lower energy—a cooling effect.

Combining these factors, researchers predicted variations of approximately one part in one hundred thousand in the temperatures of different parts of the sky (spread out by about 10 degrees in angle). The Horn antenna used by Penzias and Wilson was not sensitive enough to pick up these tiny discrepancies. But that didn't mean that these fluctuations weren't there. Radio instruments needed to catch up with the growing desire of astronomers to probe the regions of the spectrum formerly dismissed as mere static.

In the late 1970s, progress was made when Lawrence Berkeley National Laboratory (LBL) researchers George Smoot, Marc V. Gorenstein, and Richard A. Muller collected data with a more sensitive device called a differential microwave radiometer (DMR). Placed aboard a NASA Ames U-2 aircraft modified to have an opening at the top, the DMR was flown high in the atmosphere, where it collected microwave data from all over the sky in the Northern Hemisphere. A similar mission flew in the Southern Hemisphere. When the researchers plotted the CMB results, they found clear evidence of minuscule temperature differences reflecting the motion of Earth and the solar system more than 200 miles per second through space. Following the Doppler effect, photons hitting us from the forward direction of our motion were very slightly blueshifted (squeezed in wavelength) and those reaching us from behind were very slightly redshifted (stretched in wavelength), making the former a tiny bit "hotter" than the latter. "Hotter" is relative, given that the average was found to be 2.728 degrees Kelvin (above absolute zero). The researchers determined the temperature difference to be only 3.5 thousandths of a degree above or below this average. These results, along with CMB data collected by other teams throughout the 1970s using balloon-based and ground-based experiments, demonstrated conclusively that the CMB has a dipole anisotropy. "Dipole" means "having two extremes," and "anisotropy" signifies "a difference that depends upon direction." What Smoot and the others found is a minute temperature difference between the front and rear directions relative to Earth's motion, like running with warm sunshine on your face and a cool breeze blowing against your back.

Although this was a promising start, the dipole isotropy did not reveal anything about the structure of the CMB, only that we are moving through it. If you push a fork through perfectly smooth Jell-O, you could create anisotropies, too—four-pronged marks behind the fork—yet these wouldn't represent irregularities in the way the gelatinous dessert originally formed.

Various teams of astronomers, including Smoot's group at LBL, began planning for a grander experiment to map out even tinier discrepancies in the CMB that would reveal how matter was clustered during the time of its origin. Earth's atmosphere, they realized, would distort the clear signals needed. Therefore, a satellite mission would be far more promising than any aerial program for a detailed survey of the CMB. To explore space, astronomy needed to move into space and hoist its instruments into orbit. As the twentieth century approached its twilight years, a gleaming era of shiny space telescopes emerged.

Postcards from the Dawn of Time

On November 18, 1989, NASA launched the COBE (Cosmic Background Explorer) satellite, the first orbiting space probe dedicated to cosmology—specifically to mapping out the details of the CMB. Because the space shuttle program was on hiatus at the time due to the *Challenger* disaster, COBE was hoisted aboard a 116-foot-tall, revamped Delta rocket—an expendable type of launch vehicle often used to carry GPS navigational satellites and other instruments into orbit—and blasted off from Vandenberg Air Force Base near Los Angeles.

COBE contained three distinct astronomical experiments with specialized missions: the Differential Microwave Radiometers (DMR) experiment, headed by Smoot as principal investigator and astrophysicist, with Chuck Bennett of Goddard as deputy principal investigator; the Far Infrared Absolute Spectrophotometer (FIRAS) experiment, led by John Mather of NASA/Goddard Space Flight Center; and the Diffuse Infrared Background Experiment (DIRBE), steered by Mike Hauser, then also at Goddard. The DMR experiment consisted of radiometers similar to the one Smoot and his colleagues had flown aboard aircraft, tuned into three different microwave frequencies thought promising for CMB detection because there would be minimal inference from other sources, such as galaxies. Their goal was to look for minute temperature differences in the cosmic radiation between various parts of the sky that would show that the early universe was budding with structure. The radiometers were in the outer part of the satellite—just inside a protective thermal shield—where antennae picked up microwave signals directly from space and sent them to the radiometers. The information gathered was then broadcast down to Earth for interpretation.

Within the core of the satellite, enclosed in a dewar of ultracool superfluid liquid helium, were the other two experiments. FIRAS took precise measurements of the frequency distribution of the cosmic background radiation, and DIRBE looked at infrared radiation from various sources. Because of their sensitivity, they needed to be maintained at a temperature even colder than in outer space. That strict requirement set the clock ticking. The dewar-enclosed experiments could collect valid measurements only as long as the liquid helium lasted; it evaporated in less than a year. The DMR experiment was not bound by that requirement and would gather data for four years.

The COBE researchers couldn't have asked for more spectacular results. Theorists predicted that the relic radiation would have a pattern similar to a perfectly absorbing object (called a black body) cooled to 2.73 degrees Kelvin. Indeed, as the FIRAS results showed, the actual spectral curve fit like an expertly tailored suit. There was just the right amount of brightness associated with each frequency to match what would be expected for a fiery Big Bang cooled over billions of years.

It is hard enough to forecast the weather many weeks in advance, or anticipate when and where the next major earthquake will strike. These are prognostications related to relatively close and familiar things. The COBE scientists showed how is it possible to chronicle what happened billions of years ago to regions of space now billions of light-years apart. Details of cosmic history revealed themselves like the deciphered writings on an ancient parchment. It was nothing less than a revolution in our scientific understanding of the universe.

While the FIRAS experiment confirmed with unprecedented precision that there was indeed a fiery Big Bang beginning to the universe, the DMR experiment offered vital details of how the cosmos took shape in its fledgling era. Minute temperature ripples found by the DMR experiment unveiled the seeds of the structure we see today. These warmer and colder spots offered proof positive of clustering at the time the light was first emitted during the age of recombination. Like the ultrasound image of a fetus, the varied picture showed evidence of how the universe would eventually develop.

The ripples discovered by COBE indicated that at the time of recombination, the density of the fledgling cosmos varied slightly from point to point. Over the aeons, the denser areas, due to their stronger gravitational pull, would gather more and more matter. Like dough sticking to dough in the process of kneading bread, this would build up heftier and heftier loaves of material over time—leaving the sparser areas with only the crumbs. Eventually the most massive clumps of matter would ignite in the process of nuclear fusion and become the fiery predecessors of the shining objects we see today.

The spectacular COBE results transformed cosmology into a more precise science, based to a far greater extent on statistical analysis.

Astrophysicists realized that even more detailed mappings of the CMB could zoom in on finer aspects of how the universe evolved. A team, including veterans of COBE such as Dave Wilkinson, began planning a new CMB satellite, which, after his death from cancer in 2002, would come to be called the Wilkinson Microwave Anisotropy Probe (WMAP).

In the meantime, less than five months after COBE was hoisted into the heavens, another extraordinary launching of a space-based instrument provided a long-sought, crystal-clear optical perspective. The world was ready for "eyes" on the universe that would supplement the "ears" provided by COBE's antennae. By then the space shuttle program was back in action and poised to make history.

On April 24, 1990, the shuttle *Discovery*—carrying a crew of five that included the commander, Loren Shriver, and the pilot, Charlie Bolden—launched the Hubble Space Telescope into orbit, 353 miles above Earth. Continuing the legacy of the great astronomer it is named after, it was the first telescope to be wholly free of atmospheric interference and to present an unimpeded view of astronomical bodies. Along with COBE and numerous space-based astronomical instruments that followed, the $1.5 billion mission was a sign of NASA's (and the scientific world's) commitment to peer ever outward into space and ever farther back in time in the quest for cosmic knowledge.

Hubble's first images were not as sharp as hoped. To considerable embarrassment, a flaw in the shape of the telescope's 94-inch primary mirror distorted how it collected light and blurred its vision. Fortunately, another shuttle mission, launched in 1993, provided the telescope with corrective optics, rendering its vision phenomenal.

In its more than twenty years of service, Hubble has become the poster child of contemporary astronomy. Its dreamy images of wispy, colorful nebulae and other astronomical marvels adorn countless book covers, calendars, screen savers, dormitory walls, and anywhere needed to be enlivened with spectacular vistas of space. Yet its vital scientific mission greatly transcends its public role in producing stunning photos and serving as a symbol for NASA. It has offered the best evidence of the content, structure, and dynamics of the universe—in other words, what is out there, what might be missing, how everything is arranged, and where things are going. Each week, the telescope relays approximately 120 gigabytes of data down to Earth, the equivalent of many thousands of books. Analysis of the gathered information forms a critical component of modern astronomy and has revolutionized the field.

In tandem with Hubble's wealth of information in the optical and near-infrared parts of the spectrum, a number of other space telescopes and probes have explored other spectral ranges. These include the Compton Gamma-Ray Observatory, launched by NASA in 1991; the Chandra X-Ray Observatory, launched by NASA in 1999; the Spitzer Space Telescope (infrared), launched by NASA in 2003; and the XMM-Newton X-Ray Observatory and Herschel Infrared Telescope, each launched by the ESA (European Space Agency) in 2009.

Meanwhile, thanks to the advent of high-precision digital cameras and adaptive optics—systems designed to minimize the effects of atmospheric distortion—ground-based telescopes have acquired much keener vision. A triumph of twenty-first-century ground-based telescopy is the Sloan Digital Sky Survey (SDSS): a three-dimensional mapping of more than a third of the sky that began in 2000 and has continued until the present through three phases of operation. Using a dedicated 8-foot-diameter telescope at Apache Point Observatory, New Mexico, the project has located hundreds of millions of galaxies and other astronomical objects, of which nearly two million have been analyzed via their light spectra. It has offered more information about how galaxies and other astronomical objects are arranged than ever before in history. That information has been used in tandem with CMB surveys to help understand how large-scale structure has evolved in the universe.

Happy 13.75 Billionth Birthday to Space!

One of the most important recent developments in cosmology was the launching of the WMAP satellite in 2001, designed to map out the CMB in even finer detail than COBE. WMAP was hoisted into space on June 30 from Cape Canaveral in Florida aboard a Delta II rocket and placed into a special orbit called L2 (Lagrange point 2). L2 is one of five places in the Earth-Sun gravitational system that enables objects such as satellites to remain at a steady distance from those two bodies. That is because the combined gravitational forces exerted by the Earth and Sun at those points equal the amount of centripetal (enabling circling motion) force required to revolve with those bodies. Located about four times farther away from Earth than the Moon is, the advantage of L2 is a predictable, unimpeded vantage point from which to collect microwave background radiation from space. WMAP was finally retired in September 2010, offering nearly a decade's worth of extraordinary data.

The renaming of WMAP after Wilkinson turned out to be apt, given its effectiveness as a high-precision tool for cosmology in the way he had hoped. As Mather, Page, and Peebles commented in a 2003 tribute to him after his death:

> WMAP's design follows Dave's philosophy: keep it simple, but build in abundant checks for systematic errors. He was delighted with the results; the community is witnessing yet another great advance in precision tests of the relativistic cosmological model.[6]

Periodically, NASA scientists have released reports detailing the data WMAP has collected. Each of these—the three-year report released in March 2006, the five-year report released in March 2008, and the seven-year report released in March 2010—offered eye-opening revelations about the nature of the cosmos. Each has provided stronger bounds on the geometry and content of the universe and has revealed increasingly detailed information about its primal development.

A long-standing debate in cosmology has been the precise age of the universe. Before WMAP, the best guesses were pieced together through a wobbly ladder of distance estimates, producing a range of values for what is called the Hubble constant: the rate space is expanding. Astronomers traced back the current expansion to the time of the Big Bang and calculated how long ago that must have been. If space has a flat geometry (rather than a hypersphere or hyperboloid), is filled mostly with matter, and lacks a cosmological constant, then the cosmic age is two thirds divided by the Hubble constant. Because estimates of the Hubble constant varied widely and the shape of space was unclear, pre-WMAP age estimates ranged from 10 billion years (and even younger) to 18 billion years. The lower values were smaller than the calculated ages of some of the oldest known stars. This offered a paradox: how could the universe be younger than the stars it contained?

Fortunately, WMAP's wealth of data has cleared up the situation. Its "baby picture" of the universe offers a detailed portrait of the minute fluctuations in temperature from point to point in the sky, indicating its composition during the recombination era—specifically its breakdown into visible matter, dark matter, and dark energy. The shape of these ripples has revealed that the universe has a flat geometry (as predicted by the inflationary model, a leading way of understanding how the very early universe developed). By combining this bevy of information, astronomers have pinned down the age of the universe to 13.75 billion years (plus or minus 100 million years). For the first time in history, we can talk about the dawn of time with knowledge and conviction. Unless, that is, time existed before the Big Bang—more on that later!

3

How Far Away Will the Edge Get?

The Discovery of the Accelerating Universe

We live in an unusual time, perhaps the first golden age of empirical cosmology. With advancing technology, we have begun to make philosophically significant measurements. These measurements have already brought surprises. Not only is the universe accelerating, but it apparently consists primarily of mysterious substances.

—SAUL PERLMUTTER, "SUPERNOVAE, DARK ENERGY, AND THE ACCELERATING UNIVERSE," *PHYSICS TODAY* (2003)

In the history of modern physics, each generation has confronted its own enigma. The 1900s introduced light's paradoxical property of behaving like both a particle and a wave—addressed by the development of quantum mechanics. The 1930s drew the scientific community into the baffling realm of atomic nuclei. The question of how the elements built up one step at a time led to the fields of nuclear astrophysics and Big Bang cosmology. The 1960s heralded a mod world of multifarious particles with outlandish new properties—leading to the recognition that protons, neutrons, and many other types of particles are made of quarks. Today we are confronting another such challenge: the 1998 discovery of the acceleration of the universe. How science responds may produce a seismic paradigm shift as rattling as quantum mechanics.

Before 1998, cosmologists viewed the Big Bang as akin to a hurled projectile—fated to slow down under the influence of gravity. If you launched a toy rocket up into the air with a small quantity of fuel (such as what fireworks use), you'd expect it to start out fast and get slower and slower as it rose. Eventually it would "run out of steam" and fall to the ground. That's because its initial impetus would not be enough to conquer gravity. Imagine fueling it with a more powerful propellant, giving it a much more powerful blast-off akin to real rockets. Hypothetically, if it started with high enough speed, although it would still slow down as it rose higher and higher, it might just about make it into orbit. Orbital motion is a kind of balance between falling to the ground and proceeding farther outward. Fuel the rocket even more before takeoff and, although it would still slow as it ascended, theoretically it would be able to keep going and reach deep space.

Although these three possibilities have different endings, what they have in common is deceleration due to gravity. Gravity pulls things toward Earth (or toward other massive bodies), causing anything rising to slow down—unless it has an extra fuel source that enables it to accelerate. It would be very surprising if a baseball, for instance, sped up more and more after it was thrown.

Now let's think about the behavior of the universe. The Big Bang triggered an expansion of space that continues today. We witness this growth because all of the galaxies in space, except ones relatively nearby, are moving apart. Because of how gravity typically acts to clump things together and slow moving things down, astronomers assumed until almost the close of the twentieth century that the universe is decelerating. The question was whether or not the deceleration was enough to prevent the universe from expanding forever—akin to whether or not the rocket in our example would slow down enough to fall to the ground.

Gauging Cosmic Destiny

Einstein's general theory of relativity comes in two versions—with or without a factor called the "cosmological constant." Recall that his original theory, constructed more than a decade before the discovery that the universe is expanding, lacked such an extra factor. He had reluctantly introduced the term, which acts as a kind of "antigravity," in a vain attempt to stabilize a universe model he had developed and to prevent it from expanding or collapsing. Then, after learning that space is actually growing, he retracted the term. Although general relativity is simplest without a cosmological constant, recent findings about repulsive dark energy suggest that Einstein may have been right to include it—albeit for a very different reason.

Applied to the universe as a whole, general relativity describes how space expands outward from a point. If there is no cosmological constant term, this expansion slows over time because of the mutual gravitation of all of the matter. The solutions found by Friedmann in 1922 map out the three possible fates for an isotropic, homogeneous universe. Depending on whether the omega parameter (density divided by a certain critical value) is greater than 1, less than 1, or equal to 1, the universe is closed, open, or flat, respectively. In the closed case, although the universe expands for a long time, its growth keeps slowing so much because of gravity that it eventually stops and reverses itself—leading to cosmic contraction. That is like a rocket going up and then falling back to the ground. In the open case, gravity is never strong enough to halt the expansion and it continues forever, albeit more and more slowly. That is like a rocket having enough initial fuel to zoom off into deep space. Finally, the flat case involves just enough matter per volume for cosmic expansion to teeter on the brink between the two other possibilities, like a rocket launched into orbit. Effectively, that means that the universe expands indefinitely, but at a more sluggish rate than in the open case. Note that all three options, similar to the rocket example, involve deceleration—just at varying paces.

Because in general relativity the density of matter and energy is connected to the geometry of a region, there are two ways of determining which of these three fates our universe would undergo. The first method of prognosticating cosmic destiny involves adding up all the matter and radiation in space and finding out their respective densities. This is a little tricky, because in the very early universe radiation dominated over matter—meaning that it played a more important role—until, during a pivotal era tens of thousands of years after the Big Bang, the universe spread out enough for matter to take the lead. A universe that is matter-dominated expands at a somewhat faster pace than one that is radiation-dominated. Making calculations even more complicated, computing the matter density requires tallying not just the visible matter but also the dark matter that cannot be observed directly. Once all components of matter are included, comparing the actual density to a critical value divulges the universe's eventual mode of demise: "catastrophic death by too much matter" or "drawn-out death with not enough matter." In other words, the universe's density reveals its destiny, telling us if it will expire with a crunch or a whimper.

A much more direct method of gauging cosmic fate is to compare the typical recession (moving away from each other) speeds of galaxies billions of years ago and today. Galaxies' recession speeds tell us how fast the universe is expanding. Therefore, the difference in galactic recession speeds over time gauges the acceleration of the cosmos—its rate of slowing down or speeding up—and thereby tells us what will happen to space if such behavior continues. Such a calculation has become possible only with the advent of instruments able to probe the depths of observable space (the farthest places in principle that we could see). Powerful telescopes such as the Hubble Space Telescope are able to collect light from galaxies so far away that it has taken billions of years to get here. Such gathered light offers evidence from the remote past about how fast the galaxies were receding much earlier in the history of the universe.

In the 1990s, two talented teams of cosmic sleuths set out to obtain the cosmic acceleration rate—which, based upon all that was known at the time, was supposed to be a *deceleration* rate. One group, called the Supernova Cosmology Project, was based at Lawrence Berkeley National Laboratory in California and headed by Saul Perlmutter. The other, the High-Z Supernova Search, was based at Mount Stromlo Observatory in Australia and guided by team leader Brian Schmidt. Adam Riess of the Space Telescope Science Institute in Maryland was lead author on the High-Z Supernova Search team's key publications.

All evidence in cosmology until that point indicated that the expansion of the universe must be slowing. The teams wanted to know just how gradually the slowing was taking place. They raced to be the first to find out by how much cosmic growth was tapering off and thereby help determine if either a Big Crunch or a Big Whimper was in the cards. The fate of everything was in the balance.

Working independently, each team searched for Type Ia supernova explosions in distant galaxies. When a massive star undergoes a supernova burst, its core violently implodes. A colossal outpouring of energy and material is released from the star, while its interior shrinks to an incredibly dense state. Type Ia bursts, the most common type of stellar explosion, have a very standard energy profile. Like certain fast-food restaurant chains, you know exactly what you're getting no matter where a particular one is located. Therefore, as in the case of Cepheid variable stars, their regularity makes them ideal standard candles for measuring the distances to remote galaxies. By comparing their measured energies to what astronomers call their intrinsic brightness—how much energy they actually put out at the source—each team was able to calculate the distances to the supernovas they identified and the galaxies housing them.

The reason Type Ia supernovas are so regular in terms of energy output was speculated in the 1990s and would be confirmed in 2010—reported by a team led by Marat Gilfanov of the Max Planck Institute for Astrophysics in Garching, Germany. Such a burst occurs when two defunct stars, called white dwarfs, catastrophically merge. While normally a white dwarf represents the most quiescent type of stellar fate, if two of them combine they can offer an explosive coupling.

The ultimate fate of stars depends on their mass. To end up as a white dwarf, a star typically starts out at less than four times the mass of the Sun. As the star ages, it reaches a stage in which it has largely expended its hydrogen fuel and can no longer fuse it into helium. Without the outward pressure of hydrogen-burning, the helium-loaded star shrinks. This increases the internal temperature and pressure of the star, allowing helium-burning to begin. As the helium fuses into carbon, via a cycle of nucleosynthesis, the enormous quantities of radiation produced in this process force the outer envelope of the star far outward. The star becomes a red giant—a hot helium shell surrounding a dense carbon core. Eventually, though, the outer helium layers wrest free, in the colorful gaseous formation called a planetary nebula.

Once the outer material dissipates into space, the interior contracts into a small, hot object known as a white dwarf. A typical white dwarf would be only slightly larger than Earth, yet 200,000 times as dense, and have surface gravity more than 100,000 times as intense. Its interior comprises a diamondlike structure made of highly compact carbon and oxygen atoms.

White dwarfs cannot maintain their characteristic initial temperatures of more than 100,000 degrees Celsius forever. Gradually cooling over the course of billions of years, they will eventually become dark. The vast majority of stars in our galaxy—including the Sun—will end their lives in this relatively quiet way.

A much more dramatic ending could transpire if the white dwarf has a binary companion—another such object with which it is locked in a gravitational dance. Many stars belong to binary systems and other multiple-star systems. Through the decay of its orbit, a pair of white dwarfs might find itself spiraling inward and wedded into a unified object. It is a very unstable marriage because the extra mass cannot be supported by the internal pressure of the combined dwarf stars—causing a sudden implosion and ensuing release of a colossal blast of energy in a Type Ia supernova burst. Some astral couplings just aren't meant to be!

In the 1990s cosmology projects led by Perlmutter, Schmidt, and Riess, each time one of the research groups identified a Type Ia supernova, team members plotted out its energy output graph in the standard candle method to determine how far away it was. Then they applied the Doppler technique (gauging speeds through shifts in spectral lines) to find out how quickly the galaxy housing it was moving away from us. By plotting the Doppler speed data against the supernova distance information, the teams were able to determine how galactic recession rates have changed over time.

Imagine the researchers' surprise when they discovered that the universe's growth isn't slowing at all; rather, it is speeding up. Their data revealed a reckless universe slamming on the gas pedal when it was supposed to be tapping on its brakes. The astronomical community was stunned. What could be revving up the universe?

Riess described the startling moment when he realized that the universe's expansion is accelerating: "The result, if correct, meant that the assumption of my analysis was wrong. The expansion of the universe was not slowing. It was speeding up! How could that be?"[1]

Astrophysicist Michael Turner of the University of Chicago quickly recognized that none of the familiar substances in the universe could explain the extra boost the growth of space was getting. An expert on the long-standing question of what is dark matter (unseen material detected solely through its gravity), he noted that the cause of the acceleration could be a very different kind of dark constituent. Instead of clumping objects together, it pushed them apart. He dubbed the unseen repulsive agent "dark energy" and called it "the most profound mystery in all of science."[2]

In recognition of the monumental importance of their discovery, Perlmutter, Schmidt, and Riess were awarded the 2011 Nobel Prize in Physics. It was the first cosmology Nobel since 2006, when Mather and Smoot were recognized for their COBE findings about the cosmic microwave background.

Roll Over, Copernicus

Is dark energy the only option for explaining cosmic acceleration? Before concluding that any phenomenon is established fact, science must consider possible alternatives. Could what we think of as an unknown substance instead be an illusion brought about by us being in a special part of space? Might other regions have properties different from ours—mimicking the effects of dark energy? Such an alternative would require abandoning the Copernican Principle: the age-old idea that our part of the cosmos is typical rather than special.

In 2008, a thought-provoking paper, "Living in a Void: Testing the Copernican Principle with Distant Supernovae," suggested that our part of the universe might not be typical as all, but instead constitute an unusually empty region called a void. The authors, Oxford astrophysicists Timothy Clifton, Pedro Ferreira, and Kate Land, argued that setting aside the Copernican principle, as applied to the cosmological scale, could offer a natural solution to the acceleration issue without requiring dark energy. They based their arguments on earlier work by other researchers, including a Norwegian team led by Øyvind Grøn, who showed that if our part of the universe has lower than usual density, the cosmic background spectrum as recorded by WMAP would mimic that of an accelerating universe and there would be no need to postulate the existence of dark energy.

If living near the center of a giant void could mimic dark energy, how do we know which alternative is the right one? The Oxford astrophysicists proposed ways of testing that hypothesis using a refinement of the supernova standard candle technique. Although replacing dark energy with a void would eliminate one cosmological conundrum, it would have to have exactly the right properties for that substitution to work. For that reason, the paper got a mixed reaction.

As Land recalled, "Lots of people would be happy to get rid of dark energy. But the void requires almost as much fine tuning, and thus (rightly so) people were not massively optimistic. But the point of the paper was to offer up once and for all a test to separate the theories. We just needed a more comprehensive supernovae survey to decide."[3]

Suppose the Copernican Principle topples. What then? The team members pondered what would happen to the field if we were forced to rethink all our assumptions about the cosmos as a whole. As they wrote,

> Such a situation would have profound consequences for the interpretation of all cosmological observations, and would ultimately mean that we could not infer the properties of the Universe at large from what we observe locally.[4]

The researchers noted, however, that there are ways of being typical other than location. For example, cosmologist Alexander Vilinkin has argued for a "Principle of Mediocrity," asserting that terrestrial society is pretty much the norm compared to other possible civilizations in space. Perhaps there is some reason why being near the center of a cosmic bubble is a likely state of affairs for galaxies housing stars destined to support habitable planets. Just as earth-dwellers often reside in or near metropolitan areas, perhaps universe-dwellers typically happen to live near the centers of relatively empty regions.

"In this case," the Oxford researchers concluded, "finding ourselves in the center of a giant void would violate the Copernican principle, that we are not in a special place, but it may not violate the Principle of Mediocrity, that we are a 'typical' set of observers."[5]

Most scientists would be reluctant to give up the Copernican Principle unless a flood of evidence to the contrary washed their long-held suppositions away. It has proven extraordinarily important for our understanding of the universe. If other parts of space could be radically different from ours, then cosmologists would need to grapple with complex, inhomogeneous models of reality—a daunting task.

Riess took up the challenge of seeing if astronomy could rule out the philosophically troubling hypothesis that we live in an enormous void. He assembled a team called Supernova H_0 for the Equation of State (SHOES) to pin down certain cosmological parameters with greater precision, thus offering a litmus test to compare the dark energy hypothesis with competitors such as the giant bubble idea. The term "H_0" represents the current value of the Hubble constant, the expansion rate of the universe. It consists of how fast space is expanding at a certain point—typically determined by how quickly a galaxy at that point is moving away from us—divided by the distance to that location. "Equation of state" refers to the relationship between pressure and density—in that case, of dark energy. Thus two of the important goals of the project were to firm up the value of the Hubble constant and to establish the likeliest equation of state for dark energy, revealing something about its properties. By finding these quantities, they could be compared to predictions for the giant void hypothesis.

Following in Edwin Hubble's footsteps, the team used a versatile new digital camera placed on his namesake telescope. In May 2009 the space shuttle *Atlantis* blasted off for NASA's final servicing mission to the Hubble Space Telescope. The shuttle crew installed the Wide Field Camera 3 (WFC3), a high-resolution instrument that could detect light in the visible, near-infrared, and near-ultraviolet ranges—greatly enhancing the telescope's ability to deliver crystal-clear images of galaxies and other objects. Riess's team employed the WFC3 to identify six hundred Cepheid variables in nearby galaxies that host Type Ia supernovas, with the goal of combining the information from these different kinds of standard candles to establish a more precise cosmic distance ladder.

Rung by rung, the cosmic distance ladder has been built over the decades through increasingly precise observations using standard candle methods and various other techniques. It helps astronomers feel more confident about establishing how far away newly identified objects are. Using such distances, along with velocity measurement techniques such as the Doppler effect, astronomers have been able to ascertain values of the Hubble constant, the acceleration rate, and other cosmological parameters. Riess hoped that his measurements would firm up the distance ladder enough to confirm the reality of dark energy.

As Riess explained, "We are using the new camera on Hubble like a policeman's radar gun to catch the universe speeding."[6]

In January 2011, Riess and his team announced impressively precise new values for the Hubble constant and the equation of state of dark energy. The values they obtained were consistent with the basic dark energy picture and ruled out a range of inhomogenous models in which we are located at the center of a giant void. Copernicus could rest in peace; our place in space is nothing special.

"It looks more like it's dark energy that's pressing the gas pedal," Riess remarked.

While confirming that dark energy is out there, Riess's team could not explain exactly what it is. It seems that astronomers will need to badger the mysterious substance with far more questions before it reveals its true identity.

The Theory Formerly Known as Einstein's Blunder

In one segment of the classic television program *What's My Line?*, a celebrity would challenge blindfolded panelists to guess who she or he was. The panelists would phrase their questions carefully to rule out possibilities until they could narrow down the options to as few as possible. Upon the discovery of cosmic acceleration, caused by an elusive dark energy, physicists considered carefully what to ask the mystery contender. One of the leading questions they decided to pose is whether the strength of dark energy remains constant indefinitely or if it can change over time.

If dark energy always stays at the same strength, a version of the general theory of relativity that Einstein had discarded—the form that included a cosmological constant—seemed to offer a straightforward way of modeling it. After Hubble's 1929 revelation of the expanding universe, Einstein had called the introduction of the cosmological constant term his "greatest blunder." However, the discovery some seven decades later that space is stepping up its expansion offered cosmologists an incentive to restore the discarded factor. Including a small, positive cosmological constant, symbolized by the Greek letter lambda, turned out to be the simplest way of depicting a speeding-up universe. The repulsive term had suddenly become attractive!

One way of describing the effects of dark energy has to do with its pressure. As theorists have shown, the cosmological constant term is equivalent to negative pressure—the opposite of the pushing effect of positive pressure. While positive pressure squeezes things inward, negative pressure pulls them outward. Therefore dark energy is said to have negative pressure.

Imagine that a balloon is leaking and you wish to stabilize it. If you pushed on it with positive pressure, it would deflate even faster. That wouldn't do for stability. However, if magically your hands had negative pressure, you'd touch the balloon and it would cease deflating. The negative pressure of your hands would counteract the leak and stabilize the balloon. If you then touched a balloon that was already being inflated, it would inflate even faster. You'd accelerate its rate of growth.

To test whether dark energy is steady, as in a cosmological constant, numerous research teams are trying to map galactic recession in a more detailed way. In 2010, Dutch astronomer Ludovic Van Waerbeke announced results from the most comprehensive survey to date of galactic velocities. By gauging the motion of almost half a million galaxies, his group found strong evidence that dark energy permeates the universe. However, their results were not precise enough to distinguish between steady or varying forms.

Another recent study, the Cosmic Assembly Near-IR Deep Extragalactic Legacy Survey (CANDELS), has been granted an unprecedented 902 orbits worth of time on the Hubble Space Telescope to study Type Ia supernovas in 250,000 remote galaxies. By using Hubble's Advanced Camera for Surveys (ACS), along with the WFC3, to image galaxies representing the first third of galactic evolution, they hope to firm up our picture of how the expansion rate of the universe has changed over time. Headed by Sandra Faber of UC Santa Cruz, and including Riess, the group hopes that its data will help nail down whether dark energy ever wavered in its outward push. As team member

Alex Filippenko of UC Berkeley remarked, "If we find that the nature of dark energy is not changing with time then the evidence for Einstein's cosmological constant will be even greater."[7]

CANDELS's three-year run began in 2011. Already the team has released a mountain of data offering a glimpse at the childhood days of the universe. Soon we may know if an unwavering cosmological constant is the best way to represent dark energy.

Thanks to WMAP results that have further supported the accelerating universe picture, astronomers have begun to include a cosmological constant, along with regular matter and cold (slow-moving) dark matter, as part of what they call the Lambda-CDM (Lambda Cold Dark Matter) "concordance model." The idea is that the universe has passed through stages in which different factors played important roles—a theory based on a concordance of measurements using various techniques and instruments.

Although, as WMAP has shown, dark energy is at present almost three quarters of the universe's composition, it has played a seminal role only in the past six billion years. In the first eight billion years or so of cosmic history, the universe was compact enough that the density of matter (mostly in the dark form) played a larger part, which fortunately permitted vast structures to form. We owe the existence of Earth, for example, to the consequences of that early, constructive phase. Once the cosmos spread out enough, dark energy kicked in as a major player, dwarfing the role of matter. The universe's accelerating power went into full throttle, thrusting galaxies away from one another at an ever-hastening pace. Will dark energy keep pulling the universe apart until the fabric of space is decimated? Our cosmic fate depends on the answer to that vital question.

Even if the universe continues to expand forever, there's a limit to how large the part of it we're potentially able to observe can get: estimated to be about 124 billion light-years across. In their 2005 paper, J. Richard Gott and his collaborators offered a prognostication for the maximum possible size of the part of the universe that could ever be observed.[8] In their estimation, anything that moves farther than 62 billion light-years away from us in any one direction, we would never be able to observe, even in the far future. Why? One can think of it as a kind of race between light getting to Earth from galaxies and the galaxies racing away from us faster and faster. The size of the observable universe is how far away galaxies are *now* that have given off light that can reach us. Because light is still streaming toward Earth from galaxies, if we wait millions of years, we would be able to see the light given off by galaxies that are now farther away. As the distant galaxies flee from us at an ever-increasing rate, eventually they would reach a point where their light could never reach us—not now and not ever. The boundary between which galaxies' light could eventually reach us and which could never reach us is the ultimate size of how much of the universe we could ever observe. As our cosmological explorations extend farther and farther into space, it is frustrating to think that they could someday bash into an iron barrier.

4

Why Does the Universe Seem So Smooth?

The Inflationary Era

When the allegorical man came calling . . .
he showed us a flat but inflatable ball.

—THEODORE SPENCER, *THE INFLATABLE GLOBE* (1948)

Although cosmic acceleration is baffling and unexpected, at least one aspect of it rings familiar. Physicists have long suspected that the universe experienced an extraordinarily brief interval of ultrarapid growth very early in its history, called the inflationary era. Unlike cosmic acceleration, which has gradually become more significant over the aeons, inflation took place in a flash. The results were sudden and explosive.

Inflation took a patch of space much smaller than a proton and blew it up to about the size of a baseball. A baseball! This baseball-size volume, about three inches in diameter (give or take an inch), was the precursor of today's observable universe. It is truly mind-boggling to consider that everything astronomers could potentially observe today—some 93 billion light-years across—could once fit easily into a catcher's mitt! Over the course of billions of years, ordinary Hubble expansion—much slower than inflation—would increase the observable universe's girth up to its present-day value. Such colossal growth—from smaller than a proton to baseball-size and eventually to a celestial sphere billions of trillions of miles across—certainly justifies the "Big" in "Big Bang."

Although we cannot peer beyond its observable region, we presume that the entire universe has grown, not just its knowable part. One thing to keep in mind is that if the universe is infinitely large now, it was infinitely large at the time of inflation, too. Infinity multiplied or divided by any finite factor is still infinity. But infinitely large things can become more compressed. Imagine an infinite city with an infinite number of houses, one per family. If the city council passed an ordinance condemning half of the houses and asking those residents to move in with their neighbors, the city, though still infinite, would become more crowded. Similarly, the universe before the time of inflation was far, far denser than it is today.

Physicists believe everything was once that close together because it explains why the observable universe looks so similar, no matter in which direction we look. The fabric of space is remarkably smooth. Recent measurements of its geometry, through WMAP and other means, have homed in on it being as flat as a pancake. Its shape couldn't be simpler.

However, on the menu of solutions to Einstein's equations of general relativity, flat pancakes are rare items. Why is general relativity, as applied to the universe, serving only its more basic type of geometry? This question, first posed by Princeton cosmologist Robert Dicke in 1969, is known as the flatness problem.

In standard terminology, cosmologists use the omega factor to describe the relative density of matter and energy in the universe compared to a critical amount. Only for omega precisely equal to 1 is the universe flat. If omega is even an iota higher than 1, the universe is closed, and if it is slightly lower, the universe is open. Given such a strict condition, theory seems to tell us that flat would be the hardest geometry to obtain. After all, if you throw darts at a board with an incredibly tiny bull's-eye, it would be rare for one to land exactly on target.

Could the universe just be *almost* flat, rather than completely flat? Indeed, scientists sometimes miss minuscule differences because of uncertainties in measurement. General relativity doesn't make near-flatness easy, though. In 1973, Stephen Hawking and C. B. Collins proved that in the standard Big Bang expansion any small deviation from flatness must grow over time. Thus even if omega billions of years ago was extraordinarily close to 1, by now it would either be huge or tiny—translating into a noticeable closed, overdense or open, underdense universe today, rather than what we actually observe.

Astronomers like to believe that the universe started with no special conditions. Like a pristine wilderness full of natural hills and valleys, densely forested regions, and barren plains, space began as a jumble of varying terrains and conditions. There is no reason to presume that its initial profile was flat and homogenous. Rather, if its nascent shape and energy distribution followed only the laws of chance, it would have begun as chaotic as a field of jagged rocks. What, then, smoothed the haphazard landscape into a panorama as level and uniform (on its largest scale) as a vast sandy beach? The original Big Bang theory offers no cosmic steamroller to achieve such leveling.

Another issue that the standard Big Bang theory fails to address, called the horizon dilemma, has to do with the near uniformity in temperature of the cosmic microwave background. In recent years, thanks to WMAP and other surveys, cosmologists have emphasized the anisotropies of the cosmic microwave background, as they have offered a treasure trove of information about how structure formed in the universe. However, we must not forget that these represent minuscule temperature differences from point to point in the sky. A much more prominent feature of the CMB is its large-scale uniformity in all directions. It is like a computer-generated model with flawless "skin" as far as the eye could see; only a technology geek would zoom in enough to see subtle blemishes in the pixilation.

The conventional Big Bang picture of steady spatial expansion can't explain this near-evenness in temperature. If you trace back the positions of widely separated points in the sky as far back as you'd like, the constant growth model indicates that they were never close enough to have been in causal communication with one another (able to exchange light signals). By the time of the era of recombination, when the light that would form the CMB was emitted, these points would already have been much too far away—some 20 million light-years across—to have evened out their temperatures. Yet when we examine the CMB we see that their temperatures are almost equal. How to explain such an astonishing near-coincidence?

It is like a teacher who is giving a difficult test in an enormous lecture hall and is nervous about students cheating. When the exam begins, she tells the students to sit three seats apart from one another—much too far to communicate by whispering or passing notes (cell phones are banned for the test). As the test continues, she gets increasingly nervous, however, and has them move four seats away from one another; then five seats, and so forth. By the end of the exam, each student is separated from his or her neighbor by a hundred seats. Yet strangely enough, all of the students turn in nearly identical papers—with similar lists of correct and incorrect answers. How did they coordinate their responses? The teacher could accuse them of cheating when they were closer, but they were never within arm's or ear's reach. Similarly, various points in the celestial horizon, according to the original Big Bang model, were never close enough to exchange photons and coordinate their temperatures. Then, as the horizon dilemma asks, how did they cheat?

For the students, perhaps they could cheat because they were once much closer together. Maybe before the exam they huddled together as one of them shared copies of part of the answer key he had found in a wastepaper basket. Similarly, could widely separated parts of the universe have somehow been much closer—packed together tightly enough to exchange photons and coordinate their temperatures—before they very rapidly spread apart?

A Burst of Creative Energy

To resolve the flatness problem, the horizon dilemma, and several other cosmological quandaries, including the absence of monopoles (single-poled magnets having only north or south) in the universe, several scientists in Russia and the United States, including Alexei Starobinsky, of the Landau Institute for Theoretical Physics in Moscow, and Alan Guth, now at MIT, proposed the idea of an early era of exponential expansion that pushed all irregularities well beyond the horizon of what we are able to observe today. Guth called this period the inflationary era. Growth during such an epoch would be much faster than steady Hubble expansion, stretching each minuscule fragment of space into a much, much larger region. Thus what we see in the sky was once a tiny, causally connected patch of a far more extensive, multifarious cosmic garment.

A surprisingly rapid expansion would require a special mechanism. Guth proposed an idea connected to supercooling. Supercooling occurs when a liquid is brought down to a temperature below its freezing point without solidifying. It is in a metastable state, meaning that the presence of a trigger (a seed crystal, for instance) could set in motion the process of crystallization—such as an airplane zooming through a cloud full of supercooled moisture and inducing an icy coating. If you've ever left an unopened bottle of water on your back porch on a very cold day, perhaps you've seen this effect. Inside the bottle, it's still water, but the act of picking it up jars it enough to turn it all to ice almost instantly. (It's a simple experiment you can try in the winter.)

According to Guth's inflationary scheme, the universe began in a state of high symmetry, called the false vacuum. By symmetry, we mean a kind of equality in which the strengths and ranges of forces are comparable and the masses of particles are the same (all massless). This symmetry could have represented the grand unification of the natural forces (aside from gravity) before they broke down into the strong, weak, and electromagnetic interactions. The strong interaction provides the glue that cements quarks together into nucleons (protons and neutrons), which unite, in turn, into atomic nuclei. The weak interaction offers the impetus for certain types of radioactive processes, such as when neutrons decay into protons, electrons, and antineutrinos. The electromagnetic interaction is the force between charged particles. While today these forces have very different strengths and properties, physicists believe that for an extraordinarily brief instant after the Big Bang they had similar qualities. Hence the quest for unification that would offer theoretical justification for Guth's inflationary ideas.

Much progress toward unity was made in the 1960s when physicists Steven Weinberg, Abdus Salam, and Sheldon Glashow, in their theory of electroweak unification, showed how a field called the Higgs boson could spontaneously break its symmetry, lend mass to the exchange particles for the weak interaction, and cause that force to separate in range and strength from electromagnetism, carried by the massless photon. Electroweak unification became seen as a model for a possible grand unification that also included the strong force.

The Higgs mechanism, named for the pivotal work of British physicist Peter Higgs, envisions that Higgs bosons originally possessed perfect gauge symmetry. A gauge, in physics, is a kind of pointer that can rotate through various angles, like the arrow in a carnival wheel of fortune. In the very early universe, when temperatures were highest, this gauge was free to point in any direction. All of these possibilities had equal energies. However, as the universe cooled, it became energetically favorable for a random value of the gauge angle to be chosen, and it fell into place. For theoretical reasons, once a single Higgs boson froze into a particular configuration, all of the others throughout the universe followed suit and locked into the same state.

Think of the planes NASA uses to simulate lack of gravity. Those planes fly on long, steep, sharply sloped paths to give passengers a few seconds of weightlessness. For a moment, there is no up or down or sideways; everything, from humans to equipment to drops of water, floats without a preferred direction. Then, all of a sudden, the plane turns, and everything has a set of new rules to obey. Up is up and down is down, and nothing is exempt.

Likewise, the Higgs boson would lose its symmetry during the cooling of the universe and acquire a preference. As it falls to a less energetic state, it would lend mass to most of the other particles in the universe, such as the exchange particles for the weak interaction, leaving only certain types, such as the photon, as massless. Meanwhile, a remnant of the Higgs would be left over as a massive scalar field.

A scalar field is one that can be described by a single value for each point in space that is independent of the coordinate system being used. It does not depend on the location or direction of the coordinate axes. We can think of a weather forecaster's temperature map as an example. If a local government suddenly decreed that north would henceforth be known as west, and all directions would similarly be rotated 90 degrees clockwise, a weather forecaster would not have to change the temperatures on a map. That is because temperature is a scalar quantity with only magnitude, but no direction.

Through the mechanism of spontaneous symmetry-breaking and the production of a scalar field, Guth saw the opportunity to describe a phase change for the primordial universe, analogous to supercooling. Plug in a scalar field to the Einstein's equations of general relativity and the solution models a universe that is exponentially growing in scale. The field could be the Higgs or any other scalar field. Cosmologists call this solution de Sitter space-time, named for the Dutch mathematician Willem de Sitter, who derived it. (Space-time means space and time combined into a single, four-dimensional entity. It is the clay with which general relativity's universe models are molded.) A scalar field's effect on space-time is akin to the burst of volume triggered by a cosmological constant.

Guth imagined that the symmetry-breaking process would occur through a kind of supercooling of the universe right after the Big Bang, in which it would be in a metastable state primed for transformation. It would be like the bottle of water left out on the frigid porch, with a temperature below freezing but still in the liquid state. As the universe continued to cool, patches of false vacuum would spontaneously lose their initial symmetry, decay into a lower energy state, and produce a scalar field. The field—called an inflaton—would trigger a brief but explosive inflationary epoch. (The term "inflaton" is used so as not to bias the theory toward a particular scenario; it could be the Higgs or another scalar field altogether. Inflation is a process; an inflaton—note the spelling—is a scalar field that drives that process.)

During an interval of about 10^{-32} seconds—more than one quadrillion times faster than ultrashort laser pulses, some of the quickest measured events—the volume of space would increase by a factor of more than 10^{78} (1 followed by seventy-eight zeros). Imagine if a grain of sand suddenly blew up to become larger than the Milky Way; that gives you an idea of the colossal burst of expansion during that fleeting instant.

What happened right after the inflationary blast would set the course of all cosmic history. In a proposed process called "reheating," enormous, locked-up quantities of energy would flood space with massive amounts of particles. Theorists have estimated that some 10^{90} (1 followed by ninety zeros) particles emerged during reheating. These would constitute the essence of the material from which the stars, planets, and everything around us would be forged over time. Thus, according to Guth's theory, it was inflation, not the initial Big Bang, that created the bulk of everything.

As Guth pointed out, such an inflationary epoch performed two critical tasks. It considerably flattened the geometry of the inflated region, resolving the flatness problem. If a round grain of sand expanded to solar system size, any region of its surface would seem flatter than the plains of Kansas. Inflation similarly cleared up the horizon dilemma by postulating that the space we observe was once a tiny, causally connected enclave. This well-connected region blew up into everything we see, explaining its overall temperature uniformity.

Guth's paper on the subject[1] was revolutionary for its time, yet it left many important questions unanswered. Precisely how did the breakdown of grand unification produce the inflaton field? Particle physicists have yet to put forth a complete grand unified theory. Why did inflation suddenly end? Could inflation ever happen again?

Moreover, as Guth recognized, a significant issue plagued the paper. If the reheating scenario generated the matter and energy for the known universe, it should have spread such material out fairly smoothly, aside from the small clumps that served as the seeds for galaxies. That is, the state of the cosmos right afterward should have been like a well-mixed batch of pudding scattered with small raisins. Unfortunately, the phase transition that Guth proposed did not produce such overall uniformity. Rather, as bubbles of true vacuum (the background state of the universe after inflation) emerged among the sea of false vacuum, their energy was trapped in the bubble walls. In essence, the cosmic pudding was stuck on the sides of the bowl, rather than being mixed throughout. If the bubbles often collided, that could help mix up the energy more evenly. However, calculations showed that such collisions would be rare. With no clear mechanism to explain how such trapped energy became stirred throughout space, Guth's theory possessed what was called a "graceful exit" problem. There seemed to be no way to explain how the hodgepodge of bubbles, with energy confined to their walls, transformed into a more homogeneous blend.

To address this issue, Russian physicist Andrei Linde proposed a variation called "new inflation." Working independently at the University of

Pennsylvania, Paul Steinhardt and Andreas Albrecht put forth a similar approach around the same time. New inflation circumvents the graceful exit problem through the use of an inflaton field rolling along a more gradual potential. A potential represents the energy level associated with a given location. Like a ski slope, it can have a steep drop, taper off more gradually, flatten out, or rise. It can even have valleys— regions bounded on both sides by hills. These are commonly known as potential wells.

When a field rolls down a potential, its potential energy converts to other forms of energy, similar to a skier speeding up as he or she descends. If the field reaches a well, what happens next depends on whether quantum randomness comes into play.

Classical and quantum physics treat potential wells very differently. While in classical physics, a field with insufficient energy cannot escape a well's bounding walls, quantum physics allows tunneling into a region on the other side. Due to the probabilistic nature of quantum mechanics, however, the rate of such emergence cannot be known with certainty. As in radioactive decay, a quantum "roll of the dice" determines how long it would take for a field to escape the potential well and break free.

Because Guth's original model relies on chance quantum tunneling for a way out of inflation, it leads to multiple bubbles popping up randomly in various locations. The length of the inflationary era is set by the time it takes for regions to tunnel from the false vacuum (a higher energy), through a potential barrier, and then onward to the true vacuum (a lower energy). Although that ensures a long enough inflationary era to be effective, it makes for a haphazard endgame.

New inflation, on the other hand, pictures a potential that allows for a classical means of escape. Instead of a well with a barrier, its potential resembles a slide that starts off fairly flat but eventually becomes steeper until it plunges to the ground. At first the field slowly rolls along the upper reaches of the potential, fueling a sufficiently long inflationary era. However, once the field dropped to a minimum, inflation would cease, reheating would transpire, space would become flooded with the matter and energy we see today, and ordinary Hubble expansion would begin. Because the entire universe would evolve in similar fashion, its energy would be evenly distributed throughout, smoothing out major inhomogeneities. That avoids the graceful exit dilemma.

One issue, though, with new inflation is what is called the "fine-tuning problem." Its potential needs to be shaped just right to produce the correct dose of exponential expansion required to smooth out the universe. Particle theorists were pressed to develop a model that had a sufficiently flat potential. The need to get initial conditions just right seemed to contradict one of the principal motivations of inflation: to explain how any possible original state of the universe ended up in the special situation (flat and almost isotropic) that we see today.

To the rescue came another model by Linde, called chaotic inflation. Chaotic inflation eliminates the need for fine-tuning by imagining a kind of survival of the fittest among random quantum fluctuations in the froth of the primordial vacuum. These fluctuations emerged because of the Heisenberg Uncertainty Principle—one of the important tenets of quantum mechanics. It prescribes an inverse relationship between the uncertainties in energy and time: the more known about one, the less known about the other. For brief enough time intervals, it mandates, energy cannot be known precisely and therefore must fluctuate. Linde imagined that these sporadic fluctuations would battle in a race to produce the fastest-growing "bubble universes."

Let's step back to the dawn of time, before inflation transpired, to see how such a competition could have taken place. In the nascent cosmos, quantum uncertainty would allow scalar fields to bubble up from the sea of space. If by chance a field in a particular region had sufficient energy, it could have triggered a blast of inflation.

Linde proved that inflation would be possible under a variety of conditions. Rather than requiring a phase transition, as in Guth's original model, or a flat potential, as in new inflation, Linde demonstrated how more general conditions could have precipitated exponential growth. With such simple criteria to be met, chaotic inflation would mean lots of contenders in the primordial arena. Therefore, among the zoo of primal particles, there must have been at least some ferocious enough to set off an inflationary ruckus. As in the law of the jungle, a struggle for survival meant that the mightiest domain pushed the others aside and came to dominate. This fastest-growing "alpha space" dwarfed the punier regions and became the heart of the universe as we know it.

In short, by making entry into the inflation race easier, Linde guaranteed that there would be able competitors and that one would win (and vanquish the original Big Bang model's dilemmas).

Since the time of the proposals of Guth, Linde, Steinhardt, and Albrecht, inflation's extraordinary power to even out the universe has led numerous other theorists to fashion a seemingly endless array of variations. Each depends on a particular field theory based on suppositions derived from particle physics. Not all of these predict exponential growth; some imagine slightly less rapid but equally effective epochs of expansion. Because of the profound connection between what goes on in the subatomic realm and the expansive behavior of space, the subject of particle cosmology has burst forth almost as explosively as the fledgling universe itself.

Building the Astronomical Beehive

Describing the overall features of the universe requires a delicate balance. On the one hand, space on the very largest scale seems as smooth as silk. Looking at any direction in the sky and applying the coarsest averaging techniques, astronomers see roughly the same number and distribution of galaxies. On the other hand, whenever astronomers zoom in a little closer, they observe ample evidence of structure: clusters of galaxies, superclusters, voids (emptier regions), filaments (stringlike arrangements), and so forth. Galaxies themselves have a characteristic range of sizes, and there is much nearly empty space between them. Like the surface of the ocean as viewed from a ship and seen below from a plane, the universe is both ragged and smooth, depending on what scale you look at it.

This seeming dichotomy is borne out in the CMB. While the earliest measurements of the cosmic background radiation noted its remarkable sameness, no matter in what direction antennae observing it (such as the Horn antenna used by Penzias and Wilson) were pointed, later, more detailed surveys such as COBE and WMAP revealed its deeper, subtler structure, reflecting the clumping of matter at the time of that radiation's release.

One of the supreme virtues of inflationary theory is predicting how structure emerged from sameness. As an unexpected bonus, not only does inflation smooth out the universe on the largest observable scale, it also explains the inhomogeneities that do exist. Inflation helps explain why material arrangements in the universe tend to have a characteristic range of sizes. Cosmologists have traced back the dimensions of astronomical structures to the size of quantum scale, tremendously enhanced by inflation's ultrapowerful magnifying glass.

The era of ultrarapid expansion wiped the slate of the universe clean, obliterating all the irregularities present beforehand. In their place were the quantum fluctuations produced by sheer chance during the inflationary period itself.

As space expanded more and more, these fluctuations blew up larger and larger. Meanwhile, new fluctuations emerged, and greatly expanded as well. The results were variations on a wide range of scales. Once the inflationary era halted, the size of these wrinkles froze in and set the relative proportions of astronomical structure to vacant space. These energy clumps would serve as centers of growth for the gravitational processes that produced chunks of matter that led to stars, galaxies, and even greater formations.

Curiously, if the inflationary era represented *precisely* exponential expansion, its medley of fluctuations would be the same on all scales. That is, if you examined a patch of space, magnified it, and mapped it out again, its basic pattern of ripples would be identical. Such a situation is called scale-invariant, an emblem of complex mathematical arrangements called fractals. It is like taking a picture of a tree, noticing its pattern of branches, zooming in closer, and noting a similar structure in each of the branches themselves. For exact fractals with complete scale invariance, no matter how much you zoom in, their overall shape would be similar.

Precise analyses of the CMB through WMAP have served as litmus tests for inflationary models. With each release of microwave background radiation data, researchers have used powerful mathematical tools to represent how the primordial ripples are structured. A key tool is a process called Fourier transformation, which converts spatial patterns of waves into precise breakdowns of the size of their components. In other words, it maps out what proportions of fluctuations are large, small, and in between.

We can understand the process of Fourier transformation through the analogy of a vast stadium full of supporters of various Olympic teams. Imagine if, before the games, fans of each country's team are asked to stand up in succession, raise an Olympic flag with their right hand, and wave it back and forth. For each group of supporters, typically the younger fans jump up first, and the older ones slowly rise (with no less enthusiasm). Often these groupings are children seated (or standing) next to their parents. A wide-angle camera records the array of fans and flags.

One hundred years later, a historian discovers a photo of the supporters in the stadium and wishes to analyze what is going on. The image appears to be a jumble of flags, some higher and others lower. However, a careful Fourier transformation breaks down the structure of the "waves" of flags. It shows that the fluctuations have large, medium, and small components. The large components have approximately the width of about a hundred people, give or take a dozen or so. The medium ones lie in the range of smaller groups of a few people. Finally, the smallest elements range only a foot or two. The historian graphs out this Fourier-transformed data and uses the breakdown of components to calculate the scale of each team's group of supporters, the size of individual families, and the range of individual hand motions, respectively. Cosmologists perform a similar analysis of CMB fluctuations to determine how wavelength (distance between peaks) depends on scale.

The biennial WMAP reports have successively homed in on the specific distribution of CMB anisotropies. Careful analysis has revealed that the relic radiation is almost, but not quite, scale-invariant. The closeness to scale invariance has been extraordinary news for inflation boosters, nearly matching one of the theory's key predictions. The only tricky part has been tweaking the model just right to match the WMAP data, and then justifying such a variation through particle theory. Many cosmologists have already optimistically included inflation as part of the standard model of how the universe formed its structure. Others are reserving judgment until all other alternatives are ruled out.

Eternity's Challenge

Because of inflationary theory's ability to resolve the flatness and horizon problems (as well as several other technical dilemmas) and explain the origin of structure, it has been embraced by most cosmologists as the standard explanation of what transpired during the nascent instants of the universe shortly after the Big Bang. Most astronomical observations have been consistent with at least some of the versions of the inflationary scenario. Nevertheless, in recent years some of the original developers of inflation have pointed out some of the philosophical quandaries it introduces because of the spawning of new bubble universes from random fluctuations. Such a process, first described by Russian cosmologist Alexander Vilenkin on the basis of Linde's chaotic inflation model, is called eternal inflation.

Eternal inflation is designated as such because the process of cosmic creation would continue forever. Vilenkin calculated that the blowing up of space would increase the size of the regions qualifying to inflate even further. Thus space would be like a balloon inflated by a magician to produce more and more bubbling segments, until it is a spiderlike array of protrusions. Only in the case of the cosmos, the bubbling up of new extensions would never cease. Our multiverse would constitute the ultimate in sprawl—an ever-increasing array of new universes springing up like suburban developments in a growing metropolis.

Folk singer Malvina Reynolds once described suburban sprawl as "little boxes [that] . . . all look just the same."[2] Eternal inflation implies that we live in a multiplex of continually effervescing spaces produced through similar mechanisms. If we are in a cookie-cutter universe, endlessly producing new versions, our sense of uniqueness—already battered by the enormous size of the space we can see—would shrink down to practically nil.

Guth has pointed out the meaninglessness of singling out any aspect of the universe in the face of eternal inflation. As he wrote, "In an eternally inflating universe, anything that can happen will happen; in fact, it will happen an infinite number of times. Thus, the question of what is possible becomes trivial—anything is possible, unless it violates some absolute conservation law."[3]

According to Guth, it would be impossible to calculate the probability of any aspect of the multiverse because there would be an infinite amount of anything. You might try to enumerate some feature, such as the percentage of universes with more than 20 percent dark matter, but since there would be an infinite quantity of those, your estimate would depend on how you did the counting. Infinity divided by infinity, after all, is an impossible-to-determine fraction. Anything you can think of could be found in innumerable other spaces—making its portion meaningless.

Thus, if you are upset that your favorite candidate lost an election or that your cherished team narrowly missed the playoffs, take heart. If the eternal inflation model is correct, there would be countless copies of Earth in various other universes in which your candidate bested his or her opponents and your team became champions. There also would be other realities in which your favorite player simultaneously became world leader and international athletic superstar (maybe even producing a hit song and winning a Nobel Prize, also at the same time). On the other hand, you probably wouldn't want to know about all the versions of Earth in which your favorite candidates, teams, and even sport of choice don't even exist. Can you imagine a world without water polo, bungee-jumping, or curling?

Although eternal inflation certainly sounds bizarre, it offers an intriguing possible resolution of the dark energy conundrum. If universes of assorted traits compete with one another in a kind of survival of the fittest, perhaps the most successful universe is the one with the cosmological constant we observe today. The gauge of success would be the ability to produce intelligent observers such as humans. Our existence would be the litmus test for viability—a concept, which we'll discuss, called the Anthropic Principle.

5

What Is Dark Energy?

Will It Tear Space Apart?

The effect that bears [Casimir's] name is concerned with a new
interpretation of the vacuum, which is still so productive that if he
were with us today, he would have received the Nobel prize.

—FRANS SARIS, INTRODUCTION TO *HAPHAZARD REALITY*
BY HENDRIK CASIMIR (2010)

The discovery of cosmic acceleration has brought Einstein's "greatest blunder," the cosmological constant term that he added and later removed from general relativity, back in vogue. The notion of dark energy as a cosmological constant is attractive in its simplicity. Modifying general relativity to include that term requires a mere tweak of its equations. Once that term is included, the speeding up of cosmic expansion follows mathematically. Cosmologists now refer to the cosmological constant as part of a concordance model matching known astronomical data.

Why, though, is the cosmological constant the precise value that it is—small, but nonzero? Why isn't it bigger or, alternatively, just zero?

The universe we know only exists in this form because that term is exactly what it is. So how did that happen?

Scientific curiosity compels us to probe further and question what the term actually means. As modern instruments such as the WMAP satellite have shown, dark energy represents more than 72 percent of everything in the universe. Its accelerating effect on galaxies is one of the greatest mysteries in science. Given such a prominent role, just saying that it is caused by a constant, and not probing further, would fail to satisfy our yearning for a complete explanation of nature. Curiosity demands a more physical solution to the dark energy riddle.

Strangely enough, nothingness itself could provide the answer. One way of conceiving the cosmological constant is to think of it as the energy of the vacuum, also known as zero-point energy. In Einstein's theory of general relativity, energy affects geometry, causing it to warp, contract, or expand. Vacuum energy could serve as a stretching agent.

How could sheer nothingness repel galaxies? Such a trick is possible because space is never truly empty. What physicists call the vacuum represents the lowest energy state without any persistent particles. However, fleeting particles constantly pop into and out of existence, like bubbles in a foamy whirlpool.

Particles materialize from the void because of the Heisenberg Uncertainty Principle (a key tenet of quantum mechanics discussed earlier). It informs us that the briefer a particle's lifetime, the less we know about its energy. For a long-lasting particle, scientists can nail down its energy very precisely. On the other hand, for a short-lived particle, its energy is fuzzy. The result is that energy can emerge from sheer nothingness as long as it vanishes back into the void within a sufficiently short time. Following Einstein's famous prescription that energy can convert into mass, this energy can take the form of particles with mass. Particles can arise from empty space, exist very briefly, and then return to the void as long as their creation and destruction don't violate other conservation laws.

One such law is conservation of charge, which states that the total charge of a closed system cannot change. It implies that charged particles must spring from the void along with their opposites, like coupled dolphins leaping out of the water. If a negative electron arises, therefore, it must be paired with its counterpart, a positive positron.

With opposite charge, but similar properties otherwise, the positron is said to be the electron's antiparticle. The electron and positron barely experience reality before mutually annihilating. Such ghostly entities, popping into and out of existence in a flash, are called virtual particles.

Because of virtual particles, the vacuum is a hearty broth full of exotic entities floating to the surface and then sinking again—something like a stirred pot of alphabet soup, with the letters representing the widest range of fundamental particles and fields. Any massive particles able to be produced and not breaking a conservation law can pop up, albeit for the most fleeting of instants. Particle-antiparticle pairs flash into and out of reality throughout every sector of the vacuum. The energy of all these virtual particles adds up, meaning that the vacuum energy is not zero.

The reason why particle physicists have come to associate vacuum energy with the cosmological constant has to do with vacuum energy's behavior. Through a phenomenon called the Casimir effect, the more you squeeze the quantum vacuum together, the lower its energy density. Like water seeking its lowest possible level, systems tend to favor changes that reduce their energy. Unlike ordinary fluids with positive pressure that resist being compressed, the quantum vacuum likes being squeezed together. Consequently it is said to have negative pressure. In general relativity, a negative pressure turns out to have the same effect as a cosmological constant. Because it involves a physical phenomenon, not just an abstract term, physicists find it more descriptive.

At first glance, the fact that the vacuum energy naturally introduces a cosmological constant sounds incredibly lucky. It would seem that the vacuum energy, acting with its negative pressure, would supply just the outward push needed to accelerate the universe. The problem is, however, that the theoretical vacuum energy density (amount per volume) is too potent—by a factor of 10^{120} (1 followed by 120 zeros) compared to the measured value! Theory offers a cosmological constant immensely larger than the amount needed to explain cosmic acceleration. Therefore, one is left with the tricky dilemma of explaining why the cosmological constant is so small compared to theoretical predictions—but still not zero.

It is like seeing a neighbor pushing what looks like a hand-powered lawn mower. The lawn mower is rolling along a flat lawn with a constant speed for a time until it reaches a hill. You see the neighbor push a button, hear a slight humming sound, and the lawn mower gradually accelerates up the slope. Thinking that a small engine must be supplying the extra oomph, you ask the neighbor what is powering it. He earnestly responds: a nuclear generator. Once you hear that, your take on the situation completely changes. Instead of speculating about what kind of tiny engine is driving the piddling acceleration, you scratch your head wondering how the enormous power of a nuclear generator could be channeled and used to such minor effect. It is the same with vacuum energy: theoretically, it would be much too powerful a dynamo to explain the relatively subtle increases in galactic speeds found by astronomers.

The Mr. Universe Competition

To understand the nature of the vacuum, many physicists have turned to string theory, a description of the subatomic world that replaces point particles with minute, vibrating strands of energy. In recent decades string theory has vaulted into being the leading contender for a theory unifying the forces of nature: gravity, electromagnetism, and the strong and weak interactions.

String theory began as a theory of bosons, or force-carriers, representing vibrating energetic strands that convey the strong force. Particles fall into two basic spin categories: fermions and bosons, named after the great physicists who studied their statistical properties: Enrico Fermi and Satyendranath Bose. Spin is a fundamental way of describing how particles group together, how they respond to magnetic fields, and a host of other types of behavior. While fermions always must sit in different quantum states, like patrons in a theater with assigned seating, bosons may cluster into quantum states as much as they want, like concertgoers packed into a standing-room-only alcove. The constituents of matter, such as quarks and electrons, are fermions; the carriers of force, such as photons, are bosons. Thus the conveyors of the strong force, now known as gluons, are bosons.

The strong force has the property of confinement—meaning keeping subatomic particles close together—and strings seemed an adept way of representing such bonds. Originally strings did not model the fermions—the matter particles themselves. However, theorists soon discovered that the transformation of supersymmetry offered a way of transforming bosons into fermions; thus the latter could similarly be modeled with strings.

Supersymmetry involves the notion that every particle with one spin type has a doppelgänger with the opposite spin type. It offers a "magical" way to transform bosons into fermions and the other way around. It posits that quarks and electrons have bosonic mates called squarks and selectrons; that photons have fermionic compadres called photinos; and that, in fact, there is a match made in heaven for every lone particle. (Squarks, selectrons, and photinos are hypothetical particles with opposite spin types to quarks, electrons, and photons, respectively.)

Auspiciously, as shown by American physicist John Schwarz and French physicist Joel Scherk, supersymmetry, as applied to string theory, offers a natural way of explaining how gravitons, or gravity particles, came to be. They showed how gravitons emerge from the mathematics of the theory. Because of the natural inclusion of gravity, they emphasized how string theory should not be seen as just a theory of the strong force but also as a potential way of uniting all of the natural forces.

Schwarz, British theorist Michael Green, and others have demonstrated how the properties of elementary particles could be modeled through various modes of string vibrations. Like cello strings, superstrings can be tuned to various amounts of tension. Depending on this tension, they produce an array of different harmonic frequencies corresponding to different masses and other features seen in the particle world. Thus superstrings offer the flexibility needed to describe a full range of subatomic interactions and traits.

String theorists have developed competing explanations for how the vacuum energy cancels out—becomes close to, but not exactly, zero—and leads to the cosmic acceleration that we see. One possibility is that supersymmetry helps reduce the vacuum energy. If each boson in nature is balanced by a fermion counterpart, and vice versa, their vacuum energy contributions could at least partially negate each other, like two well-matched politicians facing off in a close election and one eking out a slim victory. In some cases these terms exactly cancel each other out, making the theoretical cosmological constant precisely zero.

A zero cosmological constant was perfect for the olden days (pre-1998), before the discovery of cosmic acceleration. However, to supply the right amount of dark energy requires a smidgen more. Therefore, unless the cancellations between boson and fermion terms were right on the mark, such a method would resolve the problem only to a limited extent—wrongly leaving theoretical predictions for vacuum energy either exactly zero or much higher than the detected value. Moreover, until experimental physicists actually detect supersymmetric companion particles—in collider debris, for example—the existence of supersymmetry is hypothetical.

Another option is a kind of "survival of the fittest" among possible vacuum states. String theory is blessed—or cursed, depending on one's viewpoint—with myriad possibilities for its vacuum configuration. Estimates are that there are at least10^{500} (1 followed by 500 zeros) types of vacuum states, each with its own twisted kind of higher-dimensional geometry, known as a Calabi-Yau manifold. Clearly, if string theory is true, the real physical universe is based on only one of these forms. But what kind of mechanism picks the winner? Is it just a random lottery, or are there selection rules?

Physicist Lee Smolin, currently at the Perimeter Institute in Waterloo, Ontario, Canada, first suggested, in a completely different context, the idea that evolution could be applied to cosmology, with various alternative universes having greater or lesser degrees of viability. Theorist Leonard Susskind has applied such a Darwinian notion to string theory, suggesting that each possible string vacuum state represents a point on a vast fitness landscape. We can imagine this as a varied terrain replete with mountains and valleys representing different fitness levels. The fittest universe would find its way to the highest peak, leaving its rivals panting below.

The idea of universes competing with one another implies that each is a member of a larger structure, the multiverse. A natural way of modeling this is through eternal inflation, the idea that primordial space was a spawning ground for bubble universes. The source of these fledgling universes was an entity called a scalar field (also called an inflaton) that had random energy values in different parts of space. Recall that a scalar field is something like a temperature map, assigning a number to each point—in the cosmology case, energy instead of degrees. In various places the field would offer the right value to trigger an ultrarapid expansion of space that would lead to bubble universes of different properties—all part of the multiverse. Some of these would remain puny, while others would inflate and become robust. Each would be marked with a fitness tag, depending on the value of its cosmological constant and other parameters.

Like an eager athlete, the actual universe would tend toward the state of highest fitness. What would reward Mr. Universe for his high achievement of being the absolute fittest? Vanity, it seems. His pinnacle of achievement is being celebrated by lots of fans—billions on Earth alone. For if the universe fails—by possessing qualities deadly to life—there would be no living, conscious beings around to celebrate its success.

Any cosmological argument based on whether people exist relies on what is called the Anthropic Principle. First introduced by Brandon Carter in 1973 (following a 1961 suggestion by Robert Dicke), the term refers to selecting viable universes from the set of all possibilities by screening them for conditions that lead to observers. We observe the universe; therefore only those options enabling us to do so could pass the litmus test for credible cosmologies. Ideally, this would narrow the options down to just one. Susskind used anthropic reasoning to argue that the fittest string vacuum configuration would generate the small cosmological constant that would lead to the evolution of galaxies and the rise of intelligent life on Earth.

Like a golfer seeking to beat par, the universe needs a low score to rack up the highest accolades. By selecting the vacuum configuration just right to minimize its cosmological constant, it would hit a cosmic hole in one. At least one race of intelligent observers would applaud the universe's choice of a small enough constant to enable a long period in which gravity dominates, stars and galaxies assemble themselves, planets form, and life evolves.

Not all scientists find the Anthropic Principle a credible tool. It has many detractors as well as supporters. Chief among the grumbles are that it isn't based on any verifiable set of equations and that it doesn't narrow down the possibilities enough. After all, it would potentially have to winnow some 10^{500} geometries down to a single champion. Not even Simon Cowell, the notoriously critical judge of televised music talent competitions, is that selective.

Smolin and South African cosmologist George Ellis have argued that in applying the Anthropic Principle to the vast terrain of string theory it would be too easy to skew the results toward any interpretation one would like. They also question the inclusion in that line of reasoning of parts of space that are impossible to detect. As they write:

There is a great deal of freedom in the choices that can be made here, and those assumptions determine the expectations for experiments; but they are untestable, in particular because all the other universes in the supposed ensemble are unobservable.[1]

Smolin has advised that the use of the Anthropic Principle could even play a negative role, by misleading researchers into considering it true science. Genuine science, he has pointed out, should make falsifiable predictions. The fact that life exists in our universe does not narrow down the scope of possibilities enough to offer clear scientific proof that the anthropic landscape idea is correct.

Ellis, along with his PhD student Ulrich Kirchner and collaborator W. R. Stoeger, has pointed out that the concepts of eternal inflation and the multiverse are based a quantum field theory of the vacuum that is not fully understood because of the cosmological constant issue. They emphasize the need for more objective, testable criteria to judge these theories. As they write:

The one way one might make a reasonable claim for existence of a multiverse would be if one could show its existence was a more or less inevitable consequence of well established physical laws and processes. . . . However the problem is that the proposed underlying physics has not been tested, and indeed may be untestable. . . . The issue is not just that the inflaton is not identified and its potential untested by any observational means—it is also that . . . we are assuming quantum field theory remains valid far beyond the domain where it has been tested . . . despite all the unsolved problems at the foundation of quantum theory, the divergences of quantum field theory, and the failure of that theory to provide a satisfactory resolution of the cosmological constant problem.[2]

Smolin and Ellis's arguments are emblematic of a growing division in theoretical physics between those willing to incorporate unobservable phenomena (the multiverse, unseen higher dimensions, undetectably minute strings, and so forth) into theories about why the universe is the way it is today, and those (such as Smolin and Ellis) who caution that cosmology should be based on testable assumptions. The problem is that there are no easy answers to certain fundamental questions, such as why the physical constants have particular values. Physicists have been left with the choice of waiting patiently for gathered data to produce unmistakable patterns that suggest new theories (such as how observed particle symmetries led to the notion of quarks), or accepting arguments based on honing down the set of all possibilities through selection rules such as the Anthropic Principle.

Cosmic Chameleons

Another prominent critic of anthropic reasoning is Princeton cosmologist Paul Steinhardt. He has long advocated models of the universe with potentially testable predictions. As he described his philosophy:

> I have a very narrow, pragmatic point of view about things. I just want to know if it's more powerfully predictive or not as predictive as a competitor. If it is, that's good; if not, that's bad. It's as simple as that. And I don't have a metaphysical point of view about it.[3]

Steinhardt's approach to the dark energy problem has been markedly pragmatic. Along with Robert Caldwell, now at Dartmouth, he developed an alternative to the cosmological constant explanation that would represent an actual substance, called quintessence, rather than ethereal vacuum energy. Quintessence derives its name from Aristotle's term for the fifth element, supplementing the four classical elements: earth, fire, air, and water. While the first four elements constituted tangible things, the fifth he assigned to the heavenly. In Steinhardt's scheme, baryons, leptons, photons, and dark matter would be the first four substances guiding the dynamics of the universe, and quintessence would be the fifth.

Uniquely, quintessence would be a substance with negative pressure. Unlike the cosmological constant—a fixed quantity—it could have an uneven distribution through space and evolve throughout time. Such changeability would make it a more attractive prospect for explaining why dark energy laid low billions of years ago during the pivotal era of structure formation but is now actively boosting spatial expansion.

The particular qualities of such material would be characterized by its equation of state. An equation of state describes the relationship between the pressure and density of a substance. It can be expressed by means of a factor called w, which represents the ratio of pressure over density. While ordinary matter and radiation have values of w that are greater than or equal to zero, representing positive or zero pressure, quintessence would have a negative value, signifying its negative pressure. The w factor of quintessence could conceivably vary over time and space, modeling its flexible nature.

As Steinhardt described such a hypothetical substance, "Quintessence encompasses a wide class of possibilities. It is a dynamic, time-evolving and spatially dependent form of energy with negative pressure sufficient to drive the accelerating expansion."[4]

If quintessence is everywhere, why haven't we detected it on Earth? Perhaps finding it is only a matter of time. Alternatively, maybe it has less influence here on Earth than in deep space. Could some feature of quintessence make it especially elusive close to home?

In 2003 cosmologists Justin Khoury (who had been a PhD student under Steinhardt) and Amanda Weltman, both then at Columbia University, postulated a type of quintessence called the chameleon particle. Named for its changing properties in different environments, the chameleon would have high mass in populated regions of the universe such as near Earth and low mass in relatively sparse regions such as intergalactic space. Its heft in our crowded part of the cosmos would guarantee that it interacted only weakly with other forms of matter, and at a short range. Hence it would be a hard particle to detect and wouldn't have affected prior tests of how gravity behaves in the solar system. On the other hand, in deepest space it would be lightweight enough that it could interact freely with other particles and exert its influence over a long range. It would make its presence known as a "fifth force" that induces acceleration. In short, it would have the required credentials for being stealthy close to home and highly influential far away.

Khoury described how he and Weltman developed the chameleon model:

> Our original motivation actually came from string theory, which predicts many different kinds of scalar particles that interact with ordinary matter with the same strength as gravity. The standard lore at the time was that if any of these scalars were light, the long-range force they would mediate would already have been detected in solar system tests of gravity. Following this line of logic, the scalar particles must have a large mass, and hence cannot act as dynamical dark energy.
>
> Amanda and I started thinking about a loophole in this argument, namely the possibility that the scalars only have a large mass in dense enough regions, such as our galaxy. This immediately led us to the broader idea that the properties of dark energy could vary with the environment, and thus could be testable locally.
>
> The biggest challenge was to make sure that chameleons could evade all existing laboratory and solar system tests of gravity. I remember being really stressed out that we might have missed one test that would rule out the entire idea![5]

As Khoury pointed out, testability is a great advantage of the chameleon model, compared to vacuum energy models involving a cosmological constant. "Vacuum energy only manifests itself on the largest observable scales," he noted, "whereas chameleon theories make testable predictions on much shorter scales, including in the laboratory."[6]

In 2009, physicist Aaron Chou and his team at Fermilab conducted the first experimental test for chameleon particles. The researchers fired a laser into a special steel-walled vacuum chamber called GammeV, aiming at a section with a strong magnetic field. Theoretically, the energy from the laser could potentially produce chameleon particles. These would quickly decay back into photons, leaving a characteristic light signature that could be detected and analyzed. While the experiment failed to find chameleon particles, it ruled out those of mass below a certain threshold. Thus it placed constraints on the theory—useful in developing further tests.

The Phantom Menace

The very fate of the universe—whether acceleration tears it apart—may depend on whether dark energy is represented by an ever-changing quintessence (such as chameleon particles) or a steady cosmological constant. In the former case, the universe would have hope for a reprieve, because there would be the possibility of dark energy eventually becoming weaker. Varying from era to era, the amount of quintessence could potentially decline in the future. Eventually gravity could become once again the dominant factor, leading to cosmic deceleration. A drop-off in the expansion rate would spare the universe from being torn apart.

In the latter case, on the other hand, the cosmological constant term would drive the universe to expand faster and faster, in a kind of Big Stretch. Over the aeons, all galaxies, except our neighbors, would retreat from view, leading to supreme isolation. An alternative way of modeling the same effect as a cosmological constant is a form of dark energy with an equation of state that has a w factor of precisely negative 1 for all times.

If dark energy were even more potent, as shown by Caldwell and his collaborators, the cosmos would eventually experience the Big Rip: a gruesome cosmic endgame in which dark energy would literally tear space to shreds. Caldwell refers to that persistent, powerfully repulsive version of dark energy as "phantom energy." It is characterized by an equation of state in which the w factor is less than negative 1. In other words, it would exert even more negative pressure than a cosmological constant.

If we wish to learn of our ultimate destiny, we must map out the dark energy in space and see if it varies over time. Astronomers have recently developed a powerful technique for establishing cosmic distance scales called Baryonic Acoustic Oscillations (BAO). It offers a standard ruler to measure spatial intervals out to very large distances. In doing so, it supplements standard candle methods such as Type-1a supernova energy profiles by adding reliable independent verification. By combining the standard rulers provided by BAO with the galactic speedometers offered by Doppler redshifts, researchers can map out the acceleration of the universe in extraordinary detail and search for possible variations in dark energy.

The BAO method is based on tracking ripples in the density of baryonic matter (stuff made of three quarks per particle, such as protons and neutrons) that have traveled outward since the time of recombination. Such ripples were triggered in the very early universe by the pressure of the primordial density perturbations pushing baryons away in a kind of sound wave. (We use the term "sound wave" to represent matter oscillations that move through space, not something audible.) Throughout the history of the cosmos, these waves have continued spreading out and can be observed through careful three-dimensional analysis of how galaxies are distributed through space.

Let us trace a BAO back to the nascent cosmos. It starts as a fluctuation in the plasma that makes up the very early universe. Such a wrinkle represents a hot spot with slightly greater temperature and pressure than those of its surroundings. The pressure tends to push baryons and other particles away, setting off a sound wave that changes the matter and energy distribution. That wave front is what makes up the BAO. Upon recombination, the energy is released and the matter becomes neutral atoms. Over the aeons these gravitate and serve as seeds for stars, galaxies, and other structures. Meanwhile, as space continues to expand, the BAOs present themselves as ever-growing ripples that jostle the galaxies and affect how they are situated in space. We can't actually see the galaxies responding to the BAOs—that is a slow and subtle process analogous to the gradual carving of the Grand Canyon by the Colorado River. However, modern cosmology has developed a sophisticated statistical toolkit for analyzing how galaxies are arranged and identifying the factors that are remnants of BAOs.

When cosmologists seek details about how about galaxies are distributed, they turn to the Sloan Digital Sky Survey—the premier three-dimensional mapping of the universe. In 2009, an international team of astrophysicists headed by Will Percival of the University of Portsmouth's Institute of Cosmology and Gravitation, along with Beth Reid of Princeton and Daniel Eisenstein of the University of Arizona, published a detailed statistical analysis of data from its second run, called SDSS-II, specifically observing how the BAOs have spread out over time.

As Percival related, using baryon acoustic oscillations as a standard ruler has advantages over the longer-used method of relying on type Ia supernovas. He described the benefits of BAOs:

> They are more robust as a tool for measuring the geometry of the universe. This follows from the fact that the physics is well understood for predicting the BAO signal.[7]

Like the rings of a mighty redwood, the BAOs offer an unmistakable record of how much growth took place in each era. Each period of the cosmological past is characterized by the redshift of light from galaxies of that age. The Sloan team was able to look back to a time when the wavelengths of galactic spectral lines were shifted by more than 50 percent from their ordinary (nonmoving) values, indicating look-back times of more than five billion years. By plotting the expansion rate over time, the researchers were able to ascertain the cosmic acceleration for each region and era, effectively mapping out the dark energy. (The survey has since proceeded to its third run, SDSS-III, scheduled for completion in 2014.)

Those hoping to avoid the ultimate decimation of the universe would be disheartened by the results. Rather than evolving heaps of quintessence that could diminish with time, the researchers demonstrated that dark energy is well represented by an unwavering cosmological constant. At each era in the past, as indicated by increasingly distant sky slices, the value for the cosmological constant has remained stable. There is no reason to expect, therefore, that cosmic acceleration will ever lessen. Rather, the expansion rate seems fated to speed up indefinitely—eventually leading, perhaps, to the complete fragmentation of space.

On the other hand, theorists can take heart that the concordance model that includes a cosmological constant term seems in excellent shape. As the BAO analysis is wholly independent of the supernova surveys, astronomy has produced still more evidence for the abundance of dark energy in the cosmos. Unfortunately, though we know that the cosmological constant offers a steady source of acceleration—and we know the precise value of that acceleration during a certain range of cosmic eras—that's about all we can say about it so far. Its physical cause is still unknown.

Until scientists conclusively identify the culprit for dark energy, we can expect more suspects to be pointed out. Holographic dark energy, based on a theory that all the information in the universe—or at least our region of space—is somehow coded on its exterior, represents another prominent possible cause. By "information" we refer to the simplest quantitative description of the properties of all particles and forces in the universe—something like the set of numerical data that tells a digital music player how to play a song. Some physicists, most prominently John Wheeler, have suggested that information is more fundamental than matter and energy. Just as the data bits in a computer file are able to reproduce an entire musical collection, perhaps a stream of digital information could somehow serve as the template for the entire universe. According to the holographic universe model, the code for the physical states of everything in the universe is etched on its outermost boundary—something like a surrounding CD that encodes the grand cosmic symphony. One implication for holography, that reality has a minimum length scale, could offer a natural way of capping the amount of vacuum energy in the universe and enabling it to be a more suitable candidate for dark energy. Thus the theory of holographic dark energy suggests that a deep aspect of reality itself—the way its physical features are coded—serves as the engine of cosmic acceleration.

6

Do We Live in a Hologram?

Exploring the Boundaries of Information

Holography. This is fantastic. Is incredible. You look at different
angles, at different aspects of the soul. In one holograph you
have all the information. It exists in one molecule
of the hologram.

—SALVADOR DALÍ, "INTERVIEW WITH AMEI WALLACH," *NEWSDAY* (JULY 1974)

The mystery of dark energy has inspired physicists to explore the
nature of space at its most fundamental level. To see what is forcing
the fabric of the universe to expand faster and faster, researchers are
attempting to hold up a magnifying glass to its tiniest threads with the
hope that its minutest structure will reveal the answer. At its smallest
scale, is reality seamless? Alternatively, could reality consist of minuscule
patches somehow stitched together?

We've spent time looking at the difference between how big the
universe is—possibly infinite—and how much of it we could possibly
know about—a sphere estimated to be about 93 billion light-years in

diameter. But is there a limit to how small anything can be? Like a computer monitor, is there a smallest possible resolution? Recent theories suggest that the universe does have something like pixels that represent the minimum possible size.

A fundamental smallest length, as we'll see, could offer just the impetus for the acceleration of the cosmos. This idea is called holographic dark energy.

Holographic dark energy relies on a theoretical approach called the holographic principle, proposed by Gerard 't Hooft in 1993 and applied to string theory by Leonard Susskind. Just as holograms capture three-dimensional images onto a two-dimensional film or plate—such as Salvador Dalí's famous 3-D portrait of Alice Cooper—the holographic principle states that all the information about a three-dimensional volume of space is encoded on its two-dimensional boundary. It is a generalization of a surprising result by Israeli physicist Jacob Bekenstein that the maximum information content of black holes depends on the size of their surface areas rather than on their volumes.

Into the Maelstrom

Black holes are highly compressed objects formed when the cores of massive stars catastrophically implode. They are so gravitationally intense that nothing could escape their grasp—not even light signals. Einstein's general relativity, which models gravity through the bending of space and time, predicts that black holes are so dense that they would greatly warp the space around them, creating an infinitely deep gravitational well. Anything venturing into the well would ultimately become pulverized into utter oblivion. (Don't think of these as ravenous and fast-growing like in The Blob, however. Rather, they grow only slowly with each meal and ingest only what is in their immediate vicinity, such as material drawn from companion stars.)

Fleeing a black hole is possible only for those outside the invisible boundary called its event horizon. The event horizon marks the dividing line between familiar physical principles and the bizarre realm within. Once inside, space and time become muddled, permitting motion forward toward the crushing center—a mathematical dead end called a singularity—but not backward into ordinary space. The flowing river of time transforms into a raging maelstrom, snaring unfortunate captives and hurling them toward the central singularity and nonexistence.

In the 1960s, while studying the problem of gravitational collapse, physicist John Wheeler thought at first that such a strange scenario couldn't be true. It was incredible to think that matter could simply drop off the face of reality and plunge into a bottomless abyss. After reviewing his calculations, however, he realized that such a fate was inevitable. To describe such objects succinctly, particularly their ability to prevent everything, including light, from escaping, he coined the expression "black hole," which quickly became accepted as the standard term.

As Wheeler (along with Kenneth Ford) wrote in his autobiography, black holes inform us "that space can be crumpled like a piece of paper into an infinitesimal dot, that time can be extinguished like a blown-out flame, and that the laws of physics that we regard as 'sacred,' as immutable, are anything but."[1]

To describe their obliterating nature, Wheeler proposed a well-known theorem, "black holes have no hair." Like cropped-haired soldiers virtually indistinguishable from their fellow recruits, Wheeler pointed out that black holes display little indication of their origins. Aside from three key facts—their mass, charge, and angular momentum (propensity to rotate)—they are cloaked in anonymity. Even if complex material, such as an interstellar spacecraft housing thousands of high-tech robots, were to plunge within, every aspect of its prior existence would be wiped clean. The black hole would simply gulp down the ship and gain a bit of mass.

One key aspect of this obliteration of information baffled Wheeler. He wondered how it would affect the amount of disordered energy in the universe. The second law of thermodynamics mandates that for a closed system the amount of entropy or disorder can never decrease. For example, a messy pile of wet sand couldn't spontaneously turn into an intricate sand castle—the additional orderly energy of human engineers (or kids playing that role) would be required. While no one can state with certainty that the universe is a closed system, one might reasonably expect that its total amount of disorder also couldn't decrease. However, as Wheeler pointed out, one could cheat the second law by funneling disordered material into black holes, using them as cosmic garbage disposals, and thereby upping the fraction of orderly material in the cosmos. Thus it seemed that black holes could turn thermodynamics on its head by reducing overall entropy.

In 1972, while completing his PhD work under Wheeler, Bekenstein arrived at an extraordinary solution to the dilemma. Rather than circumventing thermodynamics, he decided to extend it. Cleverly, he equated the entropy of a black hole with the area of its event horizon. Each time a black hole gobbled up a dose of disordered substances, its area would expand like a burgeoning waistline. The sum total of the entropy of material in the cosmos plus the area entropy of black holes would never decrease, thereby preserving the second law of thermodynamics.

One of the bases of Bekenstein's calculation was a subject called information theory. Bekenstein showed that the black hole area serves as an upper boundary for the maximum amount of information the black hole can store, but which is inaccessible to those on the outside. We can think of information as a way to encode the states of all particles and forces. While we normally think of information as something intangible, like everything we can know about a chair, in physics, we mean the minimum set of values needed to describe the properties of all the subatomic particles within the object, including how they interact with each other. It is like how an mpeg file—basically a string of digits—specifies a piece of music. If a unique chair is destroyed or a special music file erased, the information describing it is lost.

As the black hole gobbles up material, its associated information content becomes no longer available. Concurrently with the shrinking of available (and the growth of unavailable) information, the black hole's area expands. Thus the accumulation and destruction of information inside the black hole manifests itself as a growing surface area.

A monumental discovery in 1974 by British physicist Stephen Hawking lent credence to Bekenstein's black hole area hypothesis. Hawking brilliantly applied quantum mechanics to the analysis of black holes and showed that they would have temperatures and emit radiation. This radiation would emerge from the black hole due to a vacuum effect in the vicinity of its event horizon. Particle-antiparticle pairs leaping from the sea of nothingness as an effect of quantum uncertainty could find themselves split by the black hole's gravitational well before they could return to the froth. One member of the pair (the antiparticle, for example) might fall into the black hole while the other (the particle) escapes. The net result would be the black hole's release of a particle and its ensuing energy in a process called Hawking radiation. The gradual trickle of particles throughout the aeons would ultimately reduce the black hole's mass little by little until it completely disappeared. Another way of looking at this progression is that small amounts of mass would slip from the inside to the outside of the black hole through a process called quantum tunneling. Anything that radiates has a temperature, implying that black holes have temperatures, too. Hawking announced this unexpected result to an astonished audience at a talk titled "Black Holes Are White Hot."

Buoyed by Bekenstein and Hawking's results, the physics community came to view black holes as emblems of the fusion between quantum physics and general relativity—the minuscule and the mammoth—pointing the way toward a full theory of quantum gravity. With black holes as a model, researchers began to examine the information properties of general regions of space and their surface boundaries. Such explorations led in the 1990s to 't Hooft's formulation of the holographic principle and Susskind's subsequent extension of its concept to string theory.

Holographic Sky

The holographic principle hypothesizes that the exteriors of regions of space (black holes or the universe as a whole) embody the information of what is contained in their interiors. In other words, everything that happens anywhere within a three-dimensional volume is linked to data on the surface of that volume. Taken to its extreme, the holographic principle implies that the world surrounding us is an illusion—shadowy information rendered flesh, like the projected visage of Princess Leia in *Star Wars*.

To explore the astonishing idea that information content is located on the surface of a spatial region rather than determined via its volume, let's imagine an elementary school class that is asked to collect postcards, letters, and stamps from around the world. Each student mails letters to other schools and collects all the replies that he or she receives. At first the teacher asks each student to place the postcards and letters in boxes. The kids with more responses end up needing slightly bigger boxes. Then the teacher decides to hang the postcards and letters on the classroom walls. Most of the kids need only a small part of a wall to hold all their cards. However, the child who received the most replies ends up needing two full walls. The lesson is that if you display information on a surface rather than housing it within a volume, the surface fills up more quickly.

Now, instead of postcards on a wall, let's imagine spaceships approaching a massive black hole from all different directions, one at a time. As they reach its periphery, the astronauts display messages on the sides of their ships, like the Goodyear blimp. What would we see with a powerful enough telescope if we could read those comments?

As a highly compressed state of collapsed matter, a black hole doesn't have a real surface in the conventional sense of a place on which an astronaut could land. Yet its event horizon clearly marks the spherical boundary between its interior and exterior. Outside of the event horizon, the escape velocity of an incoming spacecraft would be less than the speed of light. With sufficient fuel and thrust, it could potentially reverse course and return to Earth before it is too late. Images of the ship would eventually reach home, albeit increasingly spread out in time as the vessel neared the horizon. That is because clocks on the ship would appear to slow compared to those on Earth—a phenomenon called gravitational time dilation. Only terrestrial observers, if they could somehow view the ship, would note such a prolonged time span. Those on board would see their ship's clock ticking at the usual pace.

When the ship reached the event horizon, no device inside would record any difference. To those on Earth, however, the vessel would appear to be frozen. That is because at the event horizon (and within), light signals would take an infinite amount of time to reach Earth. In other words, they couldn't. The frozen visage of the ship would represent the last light signals that managed to escape before the ship submerged into the abyss.

There would be no information about the astronauts once they were inside apart from their mass, charge, and angular momentum (propensity to rotate). The black hole would become very slightly more massive, pick up any extra charge the ship and astronauts happened to carry (static cling on their uniforms?), and rotate perhaps at a very slightly different rate.

Now we see why black hole information is encoded on the surfaces represented by their event horizons. The only messages that would reach us are those from the outside. We also note that with each gobbled-up spaceship, the surface would grow slightly bigger to accommodate the additional information. The maximum information content of a black hole is therefore proportional to its surface area rather than its volume.

The holographic principle, applied to cosmology, suggests generalizing this result to the universe itself. Imagine gathering all the information from within the horizon of the observable universe, the region from which we could conceivably detect signals. Beyond that horizon, the universe is a blank for us. By definition it would be impossible to see farther than the observable. In a sense, space beyond the cosmological horizon is a bit like the interior of a black hole. One important difference is that the cosmological horizon continuously grows with time as Earth can potentially collect the light from more distant reaches of the cosmos. (As we've discussed, in the far future this horizon will eventually reach an upper limit—but right now it's still growing.) In contrast, the event horizon of a black hole expands in area only after material is consumed. Setting that difference aside, if we complete the analogy between black holes and the observable universe, then the information within the observable universe is encoded within the cosmological horizon. It grows in proportion to that horizon's surface area.

Reality's Mosaic

Now let's see how the holographic principle could be used to explain the nature of dark energy. As we've discussed, one problem with representing the cosmological constant as vacuum energy is that its value would be 10^{120} times higher than what astronomers observe in the acceleration of space. One way of reducing this value significantly is to propose a minimum wavelength for how particle fields can oscillate. Because the wavelength of an oscillation is inversely related to its frequency, which in turn is connected with its energy, a minimum wavelength mandates an upper limit for its energy. The holographic principle offers a natural reason for a minimum wavelength: if every-thing within the universe is broadcast on the outside, only so much information can fit. Thus holography helps cap the value of the cos-mological constant.

To envision how this limit would work, let's consider the observable universe as made up of tiny pieces of information akin to pixels on a camera or computer screen, only much, much smaller. Borrowing a term used by Arthur Koestler in a different context, we call such mini-mal wavelength (or maximum energy) photons "holons." Holons would be the tiniest specks of information in the universe. The smallest length in physics is called the Planck length: about 6.4×10^{-34} inches. Below the Planck length, quantum effects render sizes too fuzzy to measure. To give an idea of how small that is, about a billion trillion trillion Planck lengths placed end to end would span the thickness of a strand of hair. The size of holons would be considerably larger than the Planck scale but small enough to cap the vacuum energy at a value consistent with the measured cosmological constant. Not only would light come in tiny packages—with energy doled out according to the light's fre-quency—those packages would come with a lower size limit.

While the minimum wavelength of the energetic fields is the size of the holons, the maximum wavelength is the size of the observable universe itself. The minimal and maximal wavelengths set upper and lower bounds on the vacuum energy—called the ultraviolet cutoff and the infrared cutoff, respectively. To understand these limits, let's think of the class collecting postcards, letters, and stamps. Imagine that the teacher decides to set strict ground rules on the size of the box each stu-dent collects their correspondence with and the amount of wall space they can later use on which to display it. Clearly, the box size places a maximum limit on the size of the cards and letters; they can't be big-ger than the box itself. On the other hand, there is also a lower limit mandated by the wall space requirement. Suppose a student fills up his box exclusively with tiny postage stamps. Compared to a box full of only large postcards, there would be an enormous number of stamps that could fit. Imagine the poor teacher having to mount thousands of stamps on the wall instead of dozens of postcards. The smaller stamps, collected in the same size volume as the cards, would need much more surface space because far more could fit in the box. Thousands of stamps would overwhelm the display space in a way that dozens of post-cards would not. Therefore, the requirement that everything fits on a certain part of the wall would set a minimum size for the box's contents—postcards rather than stamps. Similarly, the requirement that the universe's information fits on its surface sets a minimum size for the units that carry that information—hence the minimal wavelength.

The trickiest part of the holographic dark energy theory is proving that it has the right amount of negative pressure. In other words, if it is the real deal it must act to accelerate the universe. Interestingly, as Chinese physicist Miao Li demonstrated, if you consider the boundary of the observable universe to be the cosmological event horizon—the farthest place from which light can ever reach us, taking into account the acceleration of space—the holographic dark energy behaves like a fluid with negative pressure. Its w factor has a value close to negative 1, comparable to the effect of a cosmological constant. It makes the model a reasonable contender—assuming the holographic principle itself carries weight.

Holographic dark energy is a worthwhile idea, but one contingent on proving a rather sweeping conjecture. Researchers have yet to confirm or disprove that the information contained in the observable universe is inscribed on its boundary. Given the holographic principle's mind-boggling scope, involving the very limits of reality, it would seem a tall order to verify such a far-reaching hypothesis.

Nevertheless, physicist Craig Hogan, director of the Fermilab Center for Particle Astrophysics and professor at the University of Chicago, has taken on the challenge. He has offered a prediction based on the holographic principle that he and his colleagues plan to test through an interference experiment. He has theorized that the holographic principle, by capping the amount of information in the universe, would produce a minimum length scale for energy fields with detectable repercussions. (In our terminology, these would be holons—or minimal wavelength photons—like the minimum card size mentioned in the box analogy.) The lower boundary in wavelength would lead to a certain fuzziness, which he calls holographic noise. Drawing an analogy with data transfer limitations due to broadband capacity, he jokingly refers to this static as "Nature: The Ultimate Internet Service Provider."[2]

Hogan has conjectured that holographic noise may have already been detected in a gravitational wave experiment called GEO600 being conducted near Hanover, Germany. Like LIGO (Laser Interferometer Gravitational-Wave Observatory) in Livingston, Louisiana, and Hanford, Washington, GEO600 is an attempt to measure small oscillations in length scales due to the passage of powerful gravitational waves through Earth. These waves, a consequence of general relativity, would be produced as a by-product of powerful astronomical events, such as relatively close supernova bursts. Neither GEO600 nor LIGO has yet picked up such readings. However, when in GEO600 researchers reported an unexplained buzz plaguing their readings, Hogan suggested that the static was due to the graininess of space at its tiniest level. In other words, at that facility, holographic noise was making its broadcast debut.

Along with researcher Aaron Chou, Hogan is developing his own experiment at Fermilab, using a device called a holographic interferometer, or "holometer," to try to detect such jitters. Based on an optical process called interference, an interferometer is an instrument that splits up beams of light using mirrors and then allows the light to recombine. The merger produces characteristic patterns of bright and dark fringes that depend on the paths taken by the light waves.

We might imagine the process of interference as akin to rows of soldiers lining up after they have been divided into groups and marching along separate paths. If each row has been marching precisely in sync with the others for the same distances, when they assemble again, they should match up head to head and shoulder to shoulder. However, if they have been moving at different rates, fanning out, or otherwise marching to the beat of their own drummers, they might be out of step when they reassemble, and not line up exactly. Similarly, if in an interferometer a beam of light is split up and later recombined, whether or not the wave patterns match up peak to peak and valley to valley indicates something about its intermediate behavior. If they don't, there must be a length difference between the two paths.

A holometer is a type of interferometer designed to detect interference between two light beams due to fluctuations in their transverse (at right angles to the beam) positions. Hogan predicts that such transverse wiggles will be due to the graininess of space. The device at Fermilab will be equipped with special sensors designed to detect any shakiness because of environmental factors, which then would be taken into account. In that way the researchers will be more confident that any anomalous interference patterns they find would reflect jostling due to holographic noise.

If Hogan and his colleagues were to find such an effect, it would offer significant evidence of a minimum wavelength in nature. While that wouldn't conclusively prove the holographic principle, as there could be alternative explanations, it would represent a vital step forward in understanding the nature of space at its tiniest scale and the bounds on information in the universe. Such findings also could offer a boost for the theory of holographic dark energy by placing an upper limit on the energy content of the universe.

The Primal Quantum Sea

It is remarkable to think of reality having a minimum scale for which, if we peer closely enough, we would see something like the pixels of an electronic image. If space has a fundamental graininess, it would have manifested itself in the primordial era of the universe. In quantum physics, the tiniest scales correspond to the most energetic. Therefore the fledgling moments of time offer a natural laboratory for testing the implications of extreme conditions.

From our present-day vantage point in a mature universe with organized structures such as galaxies, let's step backward in time some 13.75 billion years to a far more chaotic age—the nascent instants after the Big Bang. Picture the start of time as a sea of possibility—a bubbling froth in which energetic fields would randomly rise to the surface, exist for a time, and then vanish into the depths.

There is much unknown about the initial state of the universe. Before a time of about 10^{-43} seconds, called the Planck time, the observable universe was compact enough that quantum rules applied. With its size below the Planck length of 6.4×10^{-34} inches, everything was as fuzzy as television static—completely probabilistic with no definitive parameters. We can only speculate about what happened before that fledgling moment.

Perhaps, as some physicists have suggested, the universe began as pure information, tantamount to a binary stream of 1s and 0s, before that numerical code somehow organized itself into particles (or strings). Maybe the initial geometry of space was itself quantum foam—a random mixture of geometric configurations—as Wheeler once advocated. If the portrait painted by string theorists is correct, perhaps the universe began in a higher-dimensional state. At any rate, we can assume that either the universe began with only the three spatial dimensions we see today, or that it started in a more complex state. In the latter case, perhaps there was a kind of phase transformation (like the freezing of liquid water into ice) that converted the initial tangle into the regular space with which we are familiar. Within a tiny fraction of a second, the geometry of the universe was frozen into the dimensionality it would maintain until this day. From that point forward, the drama of the material universe would take place on an expanding, three-dimensional stage.

Aside from geometry, another way of describing space is its topology. Topology means how things are connected. For example, a baseball and a bat have different shapes but similar topology, because if they were made of flexible dough, one could be molded into the other without cutting. On the other hand, a coffee cup has a different topology, because of the hole in the handle. Even if it were moldable, the hole would remain (unless the handle was sheered off). Similar in topology to coffee cups are doughnuts, hula hoops, picture frames, and even a sheet of cardboard with a single hole punched out.

The topology of space affects what would happen if someone could travel indefinitely in a straight line through the cosmos. Imagine a robot astronaut equipped with the incredible capability to recharge itself, rebuild itself if any of its parts wear down, travel fast enough to overtake any moving object (including galaxies), and survive for an indefinite period. (Naturally, we are speaking extremely hypothetically.) It sets out along a straight path in a fixed direction and keeps going as long and as far as it could possibly go, for many billions of years. For the simplest topology, it would just keep going, never returning to its starting point. If, in contrast, the topology were connected, like a looped strip of paper with its ends glued together (but in all three dimensions), it would eventually circumnavigate the entire universe and reach its starting place again.

A cosmic topology in which each direction of space is connected, like a three-dimensional game of Pac-Man, is called toroidal (from torus, or doughnut-shaped). If the universe had such a topology it would be finite rather than infinite. Nothing could travel forever without eventually returning to the same region. If it weren't for the rapid expansion of space, light could circumnavigate it and we would see multiple copies of the same galaxies. However, in practice, the growth of the universe would preclude anything from making the rounds; no real spaceships, now or in the foreseeable future, could possess such capabilities.

Interestingly, there is a way to measure the topology of the universe by looking deep into its past. If the universe possesses a finite, multiconnected (such as toroidal) topology, in its early stages, a light wave would be able to cross it. Like a Pac-Man character reaching the boundary of a maze, the wave would head out in one direction and return from the opposite direction. The outgoing and incoming waves would combine in a process called superposition and produce the peaked patterns known as standing waves, like plucked violin strings vibrating up and down. Such patterns could possibly be picked up in cosmic background radiation measurements.

A number of researchers, including Angelica de Oliveira-Costa of MIT and the team of Neil Cornish of Case Western University, David Spergel of Princeton University, and Glenn Starkman of the University of Maryland, have been looking for indications of multiconnected topologies. So far, however, clear signals of such interconnectedness have yet to turn up in WMAP reports and other CMB analyses. There were some hints in early WMAP data—mentioned in a 2003 paper by de Oliveira-Costa, her then husband and collaborator cosmologist Max Tegmark, and Andrew Hamilton[3]—but they ultimately didn't prove conclusive. The jury is still out, then, on whether space is more like flatbread, a doughnut, Swiss cheese, or some other strange topology—although flatbread now seems the most likely.

If eternal inflation is true, there could be many bubble universes with different geometries and topologies. Indeed, as Alan Guth has pointed out, the endless generation of bubble universes offers a kind of anarchy in which anything that could happen does happen somewhere in the multiverse. In such a cosmic labyrinth of universes, how could we establish unique properties of our own, such as flatness and smoothness?

Because of such tricky issues, as well as the need for competing explanations, in recent years Paul Steinhardt, though one of the pioneers of inflationary theory, has strongly argued for alternatives to inflation. Along with theorist Neil Turok, now executive director of the prestigious Perimeter Institute in Canada, he has developed a model known as the cyclic universe, which reproduces some of the predictions of an inflationary universe through a different mechanism. A cyclic universe relies on the idea that everything we see is within a vast three-dimensional membrane—called the "brane" for short—floating in a higher-dimensional sea called the bulk. Our brane periodically collides with another across an unseen extra dimension, argue Steinhardt and Turok, creating a source of energy from beyond that wipes the slate clean, making space smooth and regular. This kind of regeneration of the universe has come to be known as the "Big Bounce."

7

Are There Alternatives to Inflation?

Extra Dimensions and the Big Bounce

[The cyclic universe] can enable you to avoid the eternal inflation
nightmare. Some people think that eternal inflation is a virtue,
but from my point of view it is a nightmare.

—PAUL STEINHARDT, TALK AT "NEW HORIZONS IN PARTICLE COSMOLOGY:
THE INAUGURAL WORKSHOP OF THE CENTER FOR PARTICLE COSMOLOGY,"
UNIVERSITY OF PENNSYLVANIA, DECEMBER 11, 2009

If the universe is infinitely big, as observations and theories suggest, could it be infinitely old, too? Despite the failure of the Steady State hypothesis and the success of the Big Bang theory in explaining the cosmic microwave background radiation and other cosmological features, the notion of a universal beginning remains a philosophical sore point for many scientists. Shouldn't time, like space, be infinite? The debate has resurfaced in recent years with the idea of a Big Bounce— intended to supersede both the Big Bang and inflation with a cosmos of endless cycles.

It is a telling sign of the philosophical difficulties that have become associated with eternal inflation that one of the inflationary theory's key developers, Paul Steinhardt, has turned to a whole different cosmological explanation: the cyclic universe. Developed by him and Neil Turok, then at Cambridge, it replaces the idea of an inflationary stage altogether with the notion that our own universe comprises a three-dimensional membrane floating in a sea of higher dimensions and coming into periodic contact with another such membrane. It also does away with the concept that the Big Bang was a unique genesis and introduces the notion of a Big Bounce that happened repeatedly. In short, it substitutes a single timeline, involving a creation event and exponential expansion, with a cyclic exchange of energy.

There are long-standing arguments against the idea of a unique beginning of time, dating back to antiquity. Most ancient societies believed that time has no beginning or end and that history has repetitive aspects. To cite one of many examples, the Mayan culture relied on a calendar wheel that determined not just annual events, such as seasonal occurrences, but also cycles lasting many thousands of years. Therefore, despite the hype about the year 2012 allegedly representing the "Mayan apocalypse," the concept that the world will end without another cycle beginning is foreign to Mayan and most other ancient cultures. Rather, the Mayan and almost all other ancient civilizations believed in ceaseless renewal.

The notable exceptions to a belief in cycles are religions based on scriptural accounts of creation, such as Christianity, Islam, and Judaism. These faiths preach that worldly time had a single beginning in the divine act of genesis. They also share the concept of a future end-time. Given the idea of an immortal deity, this offers a dichotomy between heavenly eternal time and finite earthly time.

When Lemaître introduced the notion that would become known as the Big Bang, some nonreligious critics worried that he, as a member of the clergy, was attempting to transpose the story of Genesis onto physics. On the contrary, he was very principled and scientific, kept faith and physics separate, and was doing nothing of the sort. Still, many of those uncomfortable with the idea of a unique creation rallied around the Steady State theory's view of cosmic permanence. Ultimately, after the discovery of the cosmic microwave background demonstrated that the observable universe once was fiery and compact, virtually the entire scientific community came to accept the Big Bang theory. Still, among some thinkers, the hope for a solution that does not include a beginning lingers. As Steinhardt noted, "It's an advantage to have no beginning of time, because I think it's kind of a disturbing idea to go from 'no time' to time."[1]

Although there have been other proposed cosmologies that avoid a start to cosmic history, the cyclic universe idea and an earlier cosmology on which it was based called the ekpyrotic universe, developed by Steinhardt, Turok, Justin Khoury, then at Princeton, and Burt Ovrut of the University of Pennsylvania, were the first to make use of higher-dimensional M-theory. M-theory is an extension of string theory that includes vibrating membranes along with strings. One consequence of M-theory is that there could be another universe parallel to ours, separated from us by an extra dimension that we cannot access. Unseen higher dimensions might sound odd, but the concept dates at least as far back as the introduction of hyperspace in nineteenth-century mathematics.

The View from Flatland

In mathematics, the concept of higher dimensions beyond the familiar trio of length, width, and height is well established. Counting to three dimensions is easy, and it is natural to try to keep going. A mathematical point has no dimension, a line has one, a square (and other planar objects) has two dimensions, and a cube (and other bodies with volume) has three. Extrapolating to a four-dimensional hypercube is not too hard.

Imagine grabbing hold of a point and extending it in one direction, like stretching out a folded telescope. The result is a line segment, bracketed by a pair of points on either end. Then picture pulling down the line segment, like a window shade, until it opens into a square. The square is framed on all four sides by line segments. Now take the square and unfold it, like an accordion, in yet another direction. It becomes a cube, bounded by six squares. We have gone from two boundaries, to four, and then to six. Add two more, and we might try to envision taking eight cubes as the "surfaces" of a four-dimensional hypercube. Then ten hypercubes would form the "surfaces" of a five-dimensional polytope (higher-dimensional object) and so forth.

Although it is straightforward to continue such counting, actually picturing the higher-dimensional bodies is much harder. The three familiar dimensions are all at right angles to one another—as in the walls and floor at the corner of a room. How can we envision additional directions perpendicular to that trio? While mathematics keeps going, could nature stop at only three? Or are we simply limited in our perceptions? Art historian Linda Dalrymple Henderson has documented the extraordinary attempts by artists and other inspired individuals—Marcel Duchamp and Salvador Dalí, for example—to paint imaginative visions of the imperceptible.[2] Dalí's rendition of the fourth dimension in the painting *Cruxification (Corpus Hypercubus)* offers a visually stunning example.

Edwin Abbott brilliantly explores the question of how to conceive of higher dimensions in his novel *Flatland*, published in 1884. It imagines a completely flat world in which two dimensions is the norm. The beings of Flatland are planar geometric shapes that can see only along their own plane. They cannot conceive of the third dimension, height, because they have never experienced it. One day the protagonist, called A. Square, is visited by a talking sphere. The sphere lifts him out of his plane and shows him how his world appears from a three-dimensional perspective. The voyage is a true revelation for A. Square, who realizes that his limited vision had blocked him from realizing that space has a third dimension. Abbott strongly implies that we, in our seemingly three-dimensional universe, are similarly limited by our senses, and that a heightened perspective could reveal yet higher dimensions.

Space-time, the amalgamation of space and time that forms the fabric of relativity, has four dimensions. However, it doesn't introduce a new spatial dimension; rather, it combines the existing three dimensions of space and one of time into a unified entity. A higher dimension supplementing space and time is something more exotic.

Higher dimensions beyond space-time first appeared in physics in the early twentieth century when three European researchers—Gunnar Nordström of Finland, Theodor Kaluza of Germany, and Oskar Klein of Sweden—each tried independently to unify gravitation and electromagnetism under a five-dimensional umbrella. The idea was to develop a single set of equations explaining everything in nature. The fifth dimension, tacked onto previous theories like the addition to a house, offered more living space for unification to be completed.

Kaluza's and Klein's approaches, although derived from separate sets of assumptions, shared the feature that the fifth dimension could not be observed directly. Klein explained this aspect through an assumption that the fifth dimension is rolled up into a circle so minuscule that it could never be detected. The size would be on the order of the Planck length, less than one quadrillionth of a quadrillionth of an inch, the scale in which space becomes fuzzy. Using language developed shortly after Klein made his proposal, the uncertainty principle blurs the position of points on the circle, which is subject to quantum rules because of its extremely small size. It is like observing from an airplane window a long, round pipe on the ground, not being able to discern its width, and thinking it is just a straight line across the landscape. Unification proposals with at least one tiny, compact extra dimension are thus known as Kaluza-Klein theories.

In the 1930s, Einstein and two of his research assistants, Peter Bergmann and Valentine Bargmann, attempted to extend general relativity with a fifth dimension to incorporate electromagnetism along with gravity. Despite Einstein's genius and his assistants' extraordinary mathematical abilities and insights, their concerted efforts proved fruitless. To make matters more complicated, while they were toiling to unify two forces, other physicists were mapping out two more: what became known as the strong and weak interactions. Therefore unity would require a theory that blends a quartet of forces, not just a duet.

Supertheories to the Rescue!

Higher-dimensional unification attempts took a bit of a breather during the middle decades of the twentieth century. When Weinberg, Salam, and Glashow wed electromagnetism to the weak force in the 1960s, a conventional space-time gazebo offered sufficient room for the new couple. Many theorists soon recognized, however, that it would not be enough for a foursome. The standard model of particle physics could not accommodate a quantum description of gravity. This understanding led to a renaissance for Kaluza-Klein theories in the 1970s and 1980s through the theories of superstrings and supergravity, each of which included gravity.

Soon after superstring theory's emergence as a potential "theory of everything," questions arose about whether it possessed too much flexibility, as it included so many options. For example, strings could be open (both ends loose, like spaghetti) or closed (connected into a loop, like tortellini). Each variation offered credible theories. These were classified into various types, based on the symmetries of how they transform according to a branch of mathematics called group theory. Type I theory includes open and closed strings. Types IIA and IIB represent variations involving only closed strings. There are also two heterotic types, Heterotic-O and Heterotic-E, each involving closed strings transforming in different ways.

Heterotic is a term from biology meaning crossbreeding that offers superior qualities to the offspring. In string theory it is a hybrid of bosonic string vibrations moving in one direction and superstring vibrations moving in the opposite direction (left versus right or counter-clockwise versus clockwise) without interacting with each other. The existence of two directions nicely matches the chirality or "handedness" of nature, as seen in processes that have mirror-image variations.

In tandem with the development of string theory and supersymmetry, several groups of physicists, most prominently Sergio Ferrara, Daniel Z. Freedman, and Peter van Nieuwenhuizen of Stony Brook University, developed a supersymmetric extension of general relativity called supergravity. While the first supergravity theory was in four dimensions, in 1978 French physicists Eugene Cremmer, Bernard Julia, and Joel Scherk soon extended it into an eleven-dimensional theory to produce an array of fields that more closely matched the standard model of particle physics. As they demonstrated, the additional seven dimensions could be compactified: made unobservable by rendering them as tiny loops along the lines of Klein's proposal. In 1981, physicist Ed Witten showed that unified theories that included the symmetry groups of the standard model required at least eleven dimensions, lending a boost to that version of supergravity.

Initially, supergravity seemed to have an advantage over superstrings in that it was based on familiar point particles and was thereby more similar to existing field theories such as the electroweak model (the successful unification of electromagnetism and the weak interaction by Weinberg, Salam, and Glashow). It represented more of an extension of the long-standing concept of how particles interact, rather than a replacement of the point particle idea with energetic vibrations. On the other hand, field theories involving point particles require a special cancellation process called renormalization to eliminate infinite terms. These terms appear because point particles have infinitesimal size. Dividing by 0 (the size of a point particle) creates a problem. Try as they might, theorists could not renormalize supergravity satisfactorily. Strings, on the other hand, have finite proportions and therefore do not require renormalization. This feature allowed them to take the lead for potential theories of everything.

The first superstring revolution occurred in 1984, when Schwarz and Michael Green, then at Queen Mary College, London, proved that Type I superstring theory is mathematically sound and can be written free of certain unwanted terms called anomalies. Accordingly, prominent physicists such as Witten jumped on the superstring bandwagon, and many other theorists followed suit, launching a whole industry involving professors, postdoctoral researchers, and graduate students around the world. Soon thereafter, thanks to theorists such as David Gross, Jeffrey Harvey, Emil Martinec, and Ryan Rohm, the various types of string theory sprouted like multicolored orchids in a tropical garden.

Other physicists, such as Michael Duff, Gary Gibbons, and Paul Townsend, began to explore what would happen if one-dimensional strings were replaced with two- or higher-dimensional membranes. Instead of stringy spaghetti-like shapes, they upgraded to floppy surfaces akin to ravioli. These membrane chefs produced a savory concoction—yet appetites had to change first, in another revolution, before others would find these delectable.

How did all the string varieties relate to one another, to supergravity, and to membranes? Identifying such connections was the outcome of the second superstring revolution, also known as the M-theory revolution. The term "M-theory" was coined by Witten during a talk at a 1995 research conference, when he described how all five superstring types could be united in an eleven-dimensional theory that reduced to supergravity in its low-energy limit and that included membranes as well as strings. Witten refused to be pinned down on what the "M" stood for, suggesting "membrane," "matrix," and "magic" as possibilities. Supporters have called it the "mother of all theories," while, as Witten has conceded, critics have found it to be "murky."

While M-theory was put forth as a way of unifying string theory, it didn't narrow down its range of predictions. If anything, the possibilities encompassed by string and membrane theories have skyrocketed into mathematical mayhem. With more than 10^{500} possible vacuum states, as it is estimated, who could call string theory simple? While Susskind's landscape approach offers an attempt to explain how the fittest string theory came to evolve from the myriad possibilities, as we've seen, it has many critics who find either anthropic reasoning and/or eternal inflation unsavory in their reference to unobservable realities. Certain theorists such as Lee Smolin remain unconvinced that M-theory offers a genuine step forward, and call for a look at alternatives, such as an approach called loop quantum gravity. Nevertheless, M-theory has become a source of speculation in cosmology, particularly through its spawned concept of braneworlds.

Stringy and the Brane

In 1989, Jin Dai, R.G. Leigh, and Joseph Polchinski, working at the University of Texas at Austin, and, independently, Czech theorist Petr Hořava had found that a particular type of membrane, called a Dirichlet-brane or D-brane for short, acts as a terminus for the loose ends of open strings. It serves as a kind of flypaper, affixing both ends of open strings to its sticky wall. Closed strings, on the other hand, do not attach to D-branes and can venture away from them. Given that gravitons are represented by closed strings and that other matter and energy particles are represented by open strings, stickiness to D-branes offered a natural distinction between gravity and the other natural forces. This difference was noted in a paper by Hořava and Witten[3], which suggested that all the familiar particles, save gravitons, are attached to a D-brane. Within this D-brane they are free to move, but they cannot escape the brane itself.

Subsequently, a team of physicists from Stanford—Nima Arkani-Hamed, Savas Dimopoulos, and Georgi (Gia) Dvali (collaborating in one paper with Ignatius Antoniadas)—saw that distinction as a promising way of resolving a long-standing quandary called the hierar-chy problem. The hierarchy problem is the question of why gravity is so much weaker than the other natural forces. On the atomic scale, for example, electromagnetism and the weak interaction (the force that causes certain types of radioactive decay) are each trillions of trillions of times stronger than gravity.

It is odd to think of gravity as a weakling, especially when carrying heavy suitcases or moving furniture. Yet when we lift something we are counteracting the total gravitational force of the entire Earth. A simple experiment demonstrates gravity's relative weakness. Scatter small steel thumbtacks on a table and grab a refrigerator magnet. Then, holding the magnet above the tacks, you can vie with Earth's gravity to see which is stronger. Chances are, the small magnet will beat Earth, even with its home planet advantage.

Arkani-Hamed, Dimopoulos, and Dvali's 1998 proposal, known as the ADD model, attempts to resolve the hierarchy problem through the idea that the entire observable universe is within a three-dimensional D-brane. Another brane lies parallel to ours, merely 1/25 inch away. Represented by open strings, the particles of light and matter would be stuck on our brane and could never access the other one. Consequently we could never see or touch the parallel brane despite it being closer than the tips of our noses. Similarly, we couldn't access the space, called the bulk, between the two branes. The only particles that could access the bulk would be gravitons, as they would consist of closed strings. Free to permeate the bulk, the carriers of gravity would thus be diluted, explaining why that force is so much weaker than the others (confined to our brane).

The ADD team, along with Stanford physicist Nemanja Kaloper, fol-lowed up their original suggestion with an imaginative proposed solu-tion to the dark matter quandary—the mystery of unseen material that can be detected only through its gravitational tugs. In what they called the "Manyfold Universe," they imagined our brane as being folded up like a colossal newspaper, with the bulk representing the thin layer between the "pages." On each sheet would be galaxies and other astro-nomical objects whose presence we'd detect in two ways—through their light (optical and other types of radiation) and via their gravitational influences. Because light would need to wind its way through all the folds of our brane, but gravity could shortcut across the bulk, there would be cases of distant, unseen galaxies tugging on closer, visible material. We would chalk up such invisible influences to dark matter, whereas they would be real, shining objects that happen to reside on other folds of the brane.

Theories in which our universe happens to reside on a D-brane have come to be known as braneworlds. There have been other such suggestions since the ADD proposal. Among these have been Lisa Randall and Raman Sundrum's proposal that the bulk is a warped anti de Sitter space. With the negative cosmological constant of the bulk helping to balance the vacuum energy of the brane, the Randall-Sundrum model is a highly regarded potential resolution of the cosmological constant problem.

Cosmic Wheels

The Steinhardt-Turok cyclic universe proposal uses the braneworld notion to try to explain where the energy of the Big Bang arose. The notion that the Big Bang had a cause and that there were cycles before it has a venerable history. Earlier cyclic ideas, notably a 1931 proposal by physicist Richard Tolman, generally relied on closed cosmologies—those that would eventually recontract down to a point in a Big Crunch. Using Friedmann's closed model, without a cosmological constant, Tolman pictured a Big Bang following on the heels of a Big Crunch, triggered by the enormous energies generated in the collapse. Thus the matter and energy of the universe would be recycled. The next Big Bang would eventually end in another Big Crunch—a pattern repeated again and again for all eternity. Consequently, time would have no true beginning or end. No one would need to wonder how everything arose.

There are two major problems with the oscillatory universe model, as Tolman's scenario is called. Tolman identified the first issue himself: the buildup of entropy with each cycle. While energy itself is keen to be recycled completely, it has the wasteful habit during many processes of converting a portion of itself into unusable form. While the first law of thermodynamics mandates energy conservation, the second law guarantees that closed, natural systems cannot reduce their overall entropy. In essence, that means that things tend to wear down.

Take, for example, a painted wooden house. If its owners didn't keep it up, over time its paint would peel and its wood would rot. Shutters might fall, and windows crack and lose their shape. Its structure might start to sag and settle into the ground. Only human intervention could preserve it; nature would slowly bring it to ruin. Such are the ravages that increasing entropy brings.

According to Tolman's calculations, the universe would be no exception. With each cycle there would be an entropy buildup, rendering consecutive cycles longer and longer. Looking backward in time, that means that past oscillations were shorter and shorter, stemming back to an original creation event. In essence, therefore, like a battery-powered clock with no replacement power source, Tolman's oscillatory universe fails to tick forever and thus does not satisfy those hoping that the universe is eternal.

The second issue (that naturally Tolman, who died in 1948, could not have known about) is that, due to the measured acceleration of space, its contraction to a point is unlikely. It isn't *impossible* that dark energy will someday abate and that the galaxies will stage a multibillionth-year reunion. However, given the galaxies' observed animosity, as shown in their ever-hastening movement away from one another, it is unlikely that their superluminal social network is full of updates planning for such a coming-together event.

Given cosmic acceleration, therefore, cyclic models such as Tolman's that are based on a Big Crunch segueing into a new Big Bang are no longer physically realistic. As a result, the Steinhardt-Turok model doesn't rely on the recollapse of a closed universe to create the conditions for a new era. Rather, it imagines a flat cosmos speeding up in its expansion and continuing to cool down until it reaches empty, frigid conditions. Only then would enormous quantities of energy flood into space from a catastrophic encounter with a neighboring brane and regenerate the universe.

Fire and Ice

The first braneworld cosmology, proposed in 2001 by Steinhardt, Turok, Khoury, and Ovrut, had a rather esoteric name: the ekpyrotic universe. *Ekpyrosis*, the ancient Greek word for conflagration, was used in the Stoic tradition to describe the fiery destruction of the cosmos that marked the end of one cycle and the beginning of the next. The researchers hypothesized that the energy produced in a collision between our brane and a nearby one would transform into the familiar primordial stew of elementary particles. The universe would start off flat and homogeneous due to the smoothing character of that energy.

Slight differences in the times when the collision occurs along different parts of our brane would lead to primitive density variations. These time differences would be due to quantum fluctuations. The result would be an uneven pattern of bumps. It is like two cars colliding and scraping against each other at various random points on their exteriors as they are screeching to a halt. Each car would display uneven dents and scrapes.

For the early universe, the primordial density variations would form the gravitational seeds that would clump matter together and eventually lead to known structures such as galaxies. In short, the model would replicate the Big Bang era without an initial singularity (point of infinite density marking the beginning of time), and reproduce inflationary results without the need for a bubbling multiverse.

Khoury relates how Steinhardt, Turok, and Ovrut began to shape the rudiments of a cosmology based on branes during a railway journey in England:

> Unfortunately I wasn't part of this famous train ride, but the story as I know it is that Burt, Paul, and Neil were attending a conference in Cambridge, UK. Burt gave a lecture on his recent research, describing the possible merger of branes in M-theory and the associated particle physics. Paul and Neil both cornered Burt at the end of the lecture and asked him what would happen cosmologically if these branes collided. The "brane-storming" session occurred shortly thereafter when the three of them shared a train ride from Cambridge to London.[4]

Soon, Steinhardt and Turok realized that they could frame the brane-world collision idea in such a way that it could lead to an endless succession of cycles. The cyclic universe, as they called the extended model, would represent a clashing of our brane and the other occurring in cycles of about 1 trillion years. It would make time endless and remove the need to explain how the universe came into being.

Let's imagine the state of the universe 1 trillion years ago, before the branes collided. The energy of the interaction between the two approaching branes would accelerate the then existing galaxies and render space homogenous and flat. The cosmos would grow colder and colder and increasingly dilute. Space would stretch out more and more. Thus, before the event we call the Big Bang, the cosmos would already be smooth and regular. There would be no need for inflation afterward to do the job.

Then, 13.75 billion years ago, the mammoth collision would offer a blast of energy that would wipe out all remnants of the previous cycle—in other words, the Big Bang. The branes would recede, offering a reprieve from acceleration. New particles would be born from the collision energy, evolving into familiar nuclei, atoms, stars, galaxies, and other current astronomical objects. However, eventually the interbrane energy would kick in again, constituting a dark energy that causes galaxies to accelerate again. The pattern would repeat itself again and again.

While Steinhardt was hesitant at first about dismissing inflation theory in favor of the cyclic model, and merely called the latter an alternative, he came to see growing philosophical issues with inflation. When it was shown that inflation could sprout at any time in the history of the cosmos, producing myriad bubble universes scattered through a vast multiverse, he began to doubt its predictive value. In prepared remarks for a 2004 conference in Santa Monica, he and Turok stated,

> In models such as eternal inflation, the relative likelihood of our being in one region or another is ill-defined because there is no unique time slicing and, therefore, no unique way of assessing the number of regions or their volumes. Brave souls have begun to head down this path, but it seems likely to us to drag a beautiful science toward the darkest depths of metaphysics.[5]

At a 2009 conference in Philadelphia, Steinhardt commented that eternal inflation's inability to gauge the likelihood that the observable universe is in a certain state removes one of the key reasons why the inflationary model was originally proposed. It was supposed to predict a flat, isotropic universe, but now that it has been shown to operate eternally, producing an endless labyrinth of ever-growing bubble universes, such a prediction (or anything else that could be said about our sector of space) would be poorly defined. Referring to Guth's conjecture that "anything that can happen will happen," Steinhardt asserted that there would be no objective way of determining the probability that the observable universe possesses any particular set of conditions.

"We began with a theory that was powerfully predictive and ended up with this," he remarked. "Inflation does not explain why the universe is flat. All it has done is change the scale. I think this is serious trouble. I think that it is interesting and important that we don't have this kind of problem with the bouncing picture."[6]

Although the Steinhardt-Turok proposal has stimulated much discussion at cosmology conferences, the idea of an inflationary era has become an integral part of the narrative of how the universe developed. In particular, its prediction that quantum fluctuations, blown up enormously, formed the seeds of structure has been borne out repeatedly in astronomical surveys, including WMAP. Nevertheless, until the specific mechanism that led to an inflationary era is revealed, and the tangled skein of offshoot universes unraveled, debate about alternatives to inflation will no doubt continue.

Meanwhile, another question related to the creation of structure in the universe has been the impetus for numerous experiments. Structure formation requires sufficient matter for gravitational clumping. Yet the bulk of the needed mass cannot be visually observed, only detected through its invisible gravitational pull. What is the secret identity of that mysterious dark matter?

8

What Builds Structure in the Universe?

The Search for Dark Matter

Our existence is but a brief crack of light between
two eternities of darkness.

—VLADIMIR NABOKOV, *SPEAK, MEMORY: AN AUTOBIOGRAPHY REVISITED*

The world in which we live our daily lives is made of atoms. They make up everything from the tiniest dust mites to the tallest redwood trees, and from cups of java to molten lava. Familiar astronomical bodies—planets, stars, asteroids, and comets—are composed of atomic stuff. Even black holes are made of crushed atomic material, pulverized beyond recognition. However, atoms make up only 4.6 percent of the observable universe. The other 95.4 percent is a mystery.

WMAP results show that about 23 percent of the universe is composed of unseen matter (dwarfed only by the more than 72 percent that constitutes dark energy). Yet progress in determining what makes up that elusive dark matter has been painstakingly slow.

Swiss astronomer Fritz Zwicky discovered dark matter in the 1930s when examining the behavior of the Coma Cluster, a collection of thousands of galaxies located more than 300 million light-years away from Earth. (Recall that a light-year is about 5.9 trillion miles.) After measuring the speeds of galaxies in the cluster, he calculated how much mass it would need to have to steer them along their paths. This is similar to examining Jupiter's orbit in the solar system, calculating the gravitational force required to keep it in that trajectory, and then figuring out how massive the Sun needs to be to generate that force. While in the case of the solar system, the mass estimate would be right on target, in the case of the Coma Cluster, Zwicky found a huge discrepancy between his calculation based on galactic velocities and another he did based on how much light all the galaxies put out. Remarkably, the mass of the Coma Cluster providing its gravitational "glue" constituted 10 times the amount of mass that was radiating light. Zwicky hypothesized that most of the mass in the cluster consisted of invisible material, which he named "dark matter."

At the time of Zwicky, the study of galaxies and clusters was in its infancy. Zwicky's ideas were often controversial—he was known for being a curmudgeon—so perhaps it is not surprising that his findings raised few red flags. It took an open-minded and unassuming astronomer, Vera Rubin of the Carnegie Institution of Washington, to pick up the gauntlet in the 1970s when she discovered extraordinary results about the behavior of stars in the outer reaches of galaxies themselves. Along with Kent Ford, Rubin plotted out the speeds of stars in the outlying regions of spiral galaxies and was astonished to find that they revolved around the centers of those galaxies much faster than expected. Surprisingly, their plots of stellar speeds versus radial distances were flat, meaning that outer stars and inner stars were orbiting at the same rates. This was emphatically unlike the solar system, where outer planets, such as Neptune and Uranus revolve far more slowly than planets closer to the Sun such as Jupiter, Mars, Earth, and the inner planets. While the orbital speeds of planets in the solar system drop off with radial distance, the orbital speeds of stars in spiral galaxies remain as level as a plateau. Unseen material seemed to be giving the outer stars an extra boost.

Eerily, in recent decades astrophysicists have been able to record the ghostly influence of dark matter with greater and greater precision without knowing what it is. It is like having better and better sensors to reveal where and when a poltergeist is moving furniture around. Eventually you would hope that the poltergeist would have the manners to introduce himself.

The best means of mapping out the presence of mass involves a technique called gravitational lensing. Based on Einstein's notion that matter warps space and thus bends the paths of light rays, that method consists of looking for distortions in the image of background objects (galaxies, for instance) due to immediate bodies that might not be visible directly. Such distortions might be as obvious as double images or as subtle as a flicker in intensity as the intervening dark matter passes by. Recently, a group of astrophysicists from Japan and Taiwan applied a variation of this technique to clusters of galaxies billions of light-years away. The team demonstrated that the dark matter surrounding them is elliptical (cigar-shaped), rather than spherical, as some researchers had previously hypothesized.

Gravitational lensing has been used to study MACHOS (Massive Compact Halo Objects) in the periphery of galaxies. Such failed or extinguished stars—including faint dwarf stars and the like—may represent a certain measure of the dark material in space. However, studies have converged on the fact that unless the law of gravity is modified, a whopping portion of dark matter is in the form of unseen particles. The burning question is what they are.

A detective trying to scope out the nature of dark matter particles would have several clues with which to work. Any material that never emits light of any form must be electrically neutral. If not, electromagnetic theory tells us that charges within it would be bound under many circumstances to release photons. Much of the particle zoo flaunts its charge like a proud peacock—flashing colors (or invisible radiation, as the case may be) during motion—so the sleuth would need to focus on more bashful creatures that don't have charge. If you can't see dark matter, or otherwise detect its electromagnetic radiation, it can't be a particle that emits light of any kind.

In addition to electrical neutrality, another criterion would be a lack of response to the strong interaction. The strong force serves as a powerful binding agent keeping protons and neutrons in higher elements locked up in atomic nuclei. If dark matter particles were common and could interact via the strong interaction they would be affecting ordinary substances in discernible ways. The fact that we cannot readily detect dark matter particles anywhere here on Earth implies that they are impervious to the strong force.

That leaves two of the four natural forces: the weak interaction and gravity. A dark matter detective would be wise to home in on common, neutral particles that are weakly interacting. A leading culprit would therefore be the humble neutrino, one of the lightest known and most abundant kinds of particles. Neutrinos come in three different types, or "flavors": electron neutrinos, muon neutrinos, and tau neutrinos. As a member of the particle class called leptons, they respond to the weak interaction but not to the strong interaction. Assuming they have mass, they interact via gravity, too.

Fickle-Flavored Neutrinos

The standard model of particle physics that emerged from the successes of electroweak theory (not to be confused with the standard model of cosmology that deals with the universe itself) treats neutrinos as if they were completely massless. If that were the case they could not be gravitational sources. Massless particles cannot tug on ordinary material through their gravitational influence. Dark matter particles must have at least some mass, therefore, to perform their star-steering feats.

However, as Italian theorist Bruno Pontecorvo first conjectured, if neutrinos are able to transform from one flavor to another (which they can, in a process called neutrino oscillation), they must possess at least some mass. Given neutrinos' abundance in space, any mass they carried would add up to a hefty, but hard to detect, contribution to the material content of the universe.

Pontecorvo's life journey had many twists and turns. Born in Pisa to a Jewish family, he traveled to Rome in 1934 to work with Enrico Fermi. When the Fascist government ruling Italy at the time passed anti-Semitic laws, Pontecorvo fled to France. He went on to live in the United States and Canada, where he became involved in classified atomic projects during the Second World War. He continued his nuclear research at the Atomic Energy Research Establishment in Harwell near Oxford, England. Alarming the UK and other Western governments, he defected to the Soviet Union in 1950. Later in life he asserted that he had made the foolish choice to believe that the ideology of the Soviet government was scientifically based and would lead to a utopian society. He regretted being naive in his political judgment.

No one would dispute Pontecorvo's good judgment in the area of physics, however. A growing body of experimental evidence supports his neutrino oscillation hypothesis. In 1998, an international team of scientists from the Super-Kamiokande (Super-K) experiment based in Japan announced differences between the quantities of upward- and downward-moving muon neutrinos detected in an underground apparatus. While the upward-moving particles passed through almost the entire Earth on the way to the detector, the downward-moving particles arrived more directly from the atmosphere. The team found fewer upward-moving muon neutrinos, corresponding to theoretical predictions that a certain fraction had transformed on the way to the detector.

Not too far from where Pontecorvo once studied, a highway winds from Rome to the Adriatic coast, where sailing and other aquatic activities are popular summer pursuits. Tourists returning to the Eternal City along Highway A24 must pass through a tunnel beneath Gran Sasso, the tallest set of peaks in the Apennines. Deep within this underground passage, some may notice a special exit for a physics laboratory and wonder why a scientific facility would be in a highway tunnel. If there is a heavy traffic jam, scientists trying to leave the lab for home might be wondering the same thing.

Subterranean laboratories are ideal for neutrino detection. Thick layers of rock serve as perfect shields to block other types of particles that can't penetrate as deeply. Most such facilities are in converted mines—for example, the Homestake Gold Mine in South Dakota, where physicist Raymond Davis installed the first successful neutrino detector. That means the scientists usually need to ride down in elevators—sometimes the same ones formerly used by the miners. It is hard to take heavy equipment down. Driving to a lab is more convenient, but most cars (James Bond's aside) don't come with earth-boring devices to enable subterranean passage.

In the 1990s, the Italian government devised the ideal solution (except perhaps for the traffic). When it constructed the western-bound highway tunnel beneath Gran Sasso, it included the Gran Sasso

National Laboratory, an extensive underground physics facility, as part of the project. Neutrino researchers can experience the cosmic-ray blocking benefits of the underworld while enjoying relatively quick Fiat jaunts to the beach to relieve pasty complexions.

Since 2008, physicists at CERN near Geneva, some 450 miles away, have been lobbing Gran Sasso with volleys of particles. More specifically, CERN's Super Proton Synchrotron accelerator has been manufacturing beams of muon neutrinos and aiming them straight toward the subterranean facility. Because the neutrinos travel close to the speed of light, the trip takes only about 3 milliseconds. The bombardment is not a Swiss versus Italian squabble, nor is it a matter of jealousy about beach proximity. Rather, it involves an extraordinary experiment, called the Oscillation Project with Emulsion-tRacking Apparatus (OPERA), which would offer the first direct proof that neutrinos are able to change their flavor.

In one of the chambers within the Gran Sasso lab is an enormous detector consisting of about 150,000 light-sensitive bricks divided by lead plates and surrounded by an electronic tracking system. For three years scientists monitored the apparatus to look for signs of tau neutrinos. Billions and billions of muon neutrinos produced by CERN whizzed through the detector as if the intervening terrain were just thin air. A devastating earthquake rocked the Gran Sasso region in April 2009, killing more than 300 people in the nearby city of L'Aquila, but the lab was undamaged and the experiment continued.

Finally, in May 2010, Lucia Votano, director of Gran Sasso, announced resounding success. He and the OPERA researchers offered critical evidence that a muon neutrino had converted into a tau neutrino during its lengthy voyage. By proving that such transformations are possible, they showed that neutrinos have mass and that the standard model of particle physics is not quite right. Accordingly, they showed that neutrinos represent a viable dark matter candidate. (OPERA has since gained much press attention for a controversial 2011 claim, later cast into doubt by other groups, that the team measured neutrinos to be moving slightly faster than light.)

The dark matter mystery is far from resolved, however, because of another key qualification. The type of material needed to help create structure must be relatively slow-moving. Neutrinos flit from place to place like hummingbirds and cannot qualify as structure-builders. Rather, more lethargically paced *cold* dark matter must do the trick.

Key contenders for cold dark matter are the class of particles called WIMPs (Weakly Interacting Massive Particles). These constitute massive elementary particles that respond to the weak interaction and gravity, but not to electromagnetism and the strong force. WIMPs' insensitivity to electromagnetism—the force associated with light—explains their darkness. Unlike neutrinos, these would be heavy, slow-moving, and able to stick around long enough to serve as uniters.

Fertile ground for WIMP candidates has been cultivated through the concept of supersymmetry. Recall that this is the notion, spurred by string theory, that every particle has a companion with the opposite type of a quantum property called spin. For example, squarks—in the spin category called bosons—are the supersymmetric partners of quarks—in the spin category called fermions. If these supersymmetric companions, called superpartners, have the requisite properties, they could well be in the heavens, constituting the invisible sources of mass needed to help keep galaxies from flying apart. Whether or not superpartners exist and if they could be the long-sought WIMPs are questions that burn like phosphor in physicists' minds as they plan out searches.

WIMPs have represented the holy grail of particle cosmology, as their discovery would explain so many aspects of structure formation in the universe. Numerous teams, including two of OPERA's neighboring experiments in Gran Sasso—DAMA (DArk MAtter) and XENON100 (named after the chemical element xenon)—are searching fervently for evidence of such particles.

Will-o'-the-WIMPs

In February 2010, a team of researchers from the Cryogenic Dark Matter Search (CDMS II) experiment, in the Soudan Underground Laboratory in northern Minnesota, generated headlines and controversy with their announcement that they had identified several WIMP candidates. A former iron mine, Soudan offers almost half a mile of solid rock to shield high-energy experiments from the deluge of cosmic rays pouring through the atmosphere. Dark matter search experiments, including CDMS and the much smaller Coherent Germanium

Neutrino Technology (CoGeNT) detector, are situated in various places throughout its subterranean caverns. CDMS II features an array of drinking-glass-size germanium and silicon cylinders cooled to an incredibly frigid temperature of one-hundredth of a degree Kelvin (above absolute zero). When particles bash into one of these detectors, they generate an electric charge and deposit their energy into the material. By recording charges and energies for numerous impacts and comparing these results with the profiles expected for WIMPs, the CDMS II team found two possible contenders. Though the candidates seemed to bear some of the features expected of dark matter particles, the team acknowledged that there was something like a one-in-five chance that radioactive materials surrounding the cavern had produced signals masquerading as WIMPs.

Two weeks after the CDMS II announcement, the CoGeNT group, headed by Juan Collar of the University of Chicago, reported that the ghostly hand of dark matter might have touched their apparatus, too. Using a single germanium detector in the Soudan mine since December 2009, they found unusual energy fluctuations that could have been caused by low-mass WIMPs. In a well-publicized paper, Collar and his fellow researchers compared the "low-energy rise" they found to theoretical models (and other teams' results) and concluded:

> In view of its apparent agreement with existing WIMP models, a claim and glimmer of dark matter detection in two other experiments, it is tempting to consider a cosmological origin. Prudence and past experience prompt us to continue work to exhaust less exotic possibilities.[1]

Soon thereafter, a sharp rebuttal to these reports arrived in the form of negative results from the XENON100 experiment in Italy. Using a detector filled with more than 350 pounds of liquid xenon deep in the Gran Sasso tunnel, a team of researchers collected particle data throughout the autumn of 2009, seeking evidence of relatively low-mass WIMPs. Announced the following spring, these results were diametrically opposite to what had been reported from the Soudan mine. With 90 percent confidence, the XENON100 team determined that their detector had picked up no dark matter candidate events at all. Either dark matter's stealthy spirit preferred Minnesota mines to Apennine tunnels—which would be odd because the DAMA experiment, also in Gran Sasso, had previously found more positive indications—or a gaping inconsistency needed to be resolved.

As it stands, it looks like Canadians could serve as arbiters of this issue, as the latest batch of WIMP detection experiments have been based in Sudbury, Ontario, tucked into the deepest underground physics laboratory in the world. More than one and one-quarter miles beneath the ground, the SNOLAB (Sudbury Neutrino Observatory Lab) facility is in an active nickel mine. As its name suggests, the center was originally used for neutrino observation experiments. However, with funds from the Canadian government, it has expanded into a full-fledged haven for any group searching for hard-to-find particles from space—including WIMPs. Although researchers have to journey down a grimy mine shaft to get there, once they have reached the subterranean chamber they find it a sterile, well-protected working space separated from the noisy atmosphere by thousands of feet of granite.

In August 2010, Collar began to set up shop there for the next phase of his WIMP-hunting expedition. He and his team from the Chicagoland Observatory for Underground Particle Physics (COUPP) brought in sensitive bubble chamber detectors designed to track any stray dark matter particles through the trails they leave behind. A bubble chamber is a container of clear, superheated (teetering at a temperature just below its boiling point) fluid that exhibits vapor tracks whenever certain types of particles pass through and heat the liquid. While confident that he would be able to recognize the traces of candidate particles, he has asserted that any good experimentalist should continuously question his or her own results to make sure all other possibilities are ruled out. As Collar once noted:

> I try to teach my students that a good experimentalist does not need any critics: he or she is his/her own worst enemy. If you don't feel a sincere drive to debunk, test and revise your own conclusions, you should be doing something else for a living.[2]

Another WIMP quest, called Project In CAnada to Search for Supersymmetric Objects (PICASSO), has similarly made SNOLAB its home and has been collecting data deep underground. The CDMS group is also considering a move there—which would further establish the site as one of the premier hunting lodges for dark matter particle seekers. Because the Creighton mine where SNOLAB is housed is two-and-one-half times deeper than Soudan, it is much more effective in blocking out extraneous cosmic rays.

Meanwhile, WIMP searching has been a prime mission of particle-crashing experiments based at the world's largest accelerator, CERN's Large Hadron Collider (LHC). Researchers at the LHC hope to find evidence of massive particles that could possibly serve as dark matter components. Contenders include the lightest-mass superpartners predicted by certain supersymmetric theories. Although supersymmetry has yet to be confirmed, many high-energy physicists hope that its predicted particles will represent the first discovered cold dark matter constituents. The race is on as to whether the first bona fide WIMPs will be found among the debris of particle smashers or in the imprints of cosmic rays in underground detectors.

A Detergent for Dark Matter

WIMPs are not the only type of cold dark matter candidate. Another contender, called the axion, arose from an attempt to cleanse the standard model of particle physics and remove a marked inequity.

Physicists like to clean up theories that are stained with obvious exceptions. CP (Charge-Parity) symmetry is a rule that a particle and a mirror image of its oppositely charged version should behave in the same way. However, we do sometimes see that rule violated. The standard model of particle physics contains a mechanism for explaining CP violation in the weak interaction. That is a good thing, given that CP violation could justify why there is far more matter than antimatter in the universe today. However, those hoping for a unified theory that includes the strong force as well as the weak have been stymied by the question of why the strong interaction doesn't exhibit CP violation, too. To put this in a more positive way, if you consider quarks participating in the strong interaction by tossing gluons, change the players so the charges are opposite to that of the original, and view the match in a mirror, it would look the same. If the weak and strong interactions once were united, why did weak interaction bouts remain down and dirty, breaking CP symmetry without any qualms, while the strong interaction managed to clean up its act?

In 1977, Helen Quinn and Roberto Peccei, then both at Stanford, proposed an innovative solution to the strong CP conservation dilemma. Known as the Peccei-Quinn symmetry-breaking theory, it envisions that an extra field introduces a mathematical term into quantum chromo-dynamics (the leading model of the strong interaction) that cancels out an existing CP-violating term. Once the two terms cancel out to remove the imperfection, the strong interaction starts looking the same in a mirror, while the weak interaction keeps its blemished appearance. Physicist Frank Wilczek, then at Princeton, developed a detailed model for how the new field arises and—in honor of its ability to cleanse the theory—named it after Axion Detergent: a household laundry product.

Astrophysicists soon realized that the properties of certain types of axions, assuming they do exist, would make them ideal dark matter can-didates. For one thing, they would be electrically neutral and interact only rarely with other types of matter. Although, like neutrinos, axions would be extremely light-weight, theories of how they are formed pre-dict that they would tend to cluster together and move much more slowly than their lightness would suggest. Thus, in contrast to flighty, "hot" neutrinos, axions would be sluggish and "cold"—offering through their combined gravitational influence a sturdy type of scaffolding for the buildup of structure in the cosmos.

One possible decay mode for axions suggests a way of trying to hunt them down. Under the influence of a strong magnetic field, an axion could decay into two photons. The frequency of the photons would be proportional to their energies, which (following Einstein's famous mass-energy conversion mechanism $E = mc^2$) would depend on the axion's mass. Thus researchers looking for axions could set up a system with a strong magnetic field and look for the signature of photons pro-duced at certain frequencies. The frequencies expected for the photons associated with the decay of dark matter axions would be in the radio-to-microwave range.

The pioneering developer of the method used to search for axions is University of Florida physicist Pierre Sikivie. In 1983 he invented a device called the magnetic axion telescope, also known as a magnetic haloscope, which has become the basis of axion sleuthing. It consists of a special chamber called a resonance cavity (used to build up electromagnetic waves) surrounded by a powerful magnet. He suggested this instrument as a way of finding axions from our galaxy's dark matter halo as these featherweight particles pass through Earth.

Sikivie is now a prominent member of ADMX (Axion Dark Matter Experiment), one of the leading collaborations for trying to detect dark matter axions. The experiment took form at Lawrence Livermore National Laboratory and moved in July 2010 to the CENPA (Center for Experimental Nuclear Physics and Astrophysics) at the University of Washington.

To generate the powerful magnetic field needed to convert axions to microwave photons, ADMX uses a superconductor. Superconductors are substances that lose their electrical resistance (obstacles to current flow) at extremely cold temperatures and are able to build up high magnetic fields. Superconducting magnets are so strong, they are used for the highest-powered particle colliders and have been tested on robust magnetic levitation systems that may someday be used for trains.

The other key component of ADMX is a tunable resonance cavity that allows for standing waves of various frequencies. It is like a guitar string that can be tuned to create different vibration patterns. Imagine tuning a guitar and playing just the right note to shatter a glass vase. The guitar and vase are then said to be in resonance. The ADMX researchers are searching for the right frequency to "shatter" an axion and have it convert into microwaves. They need to adjust the input signals to find this unknown resonant frequency. Once the microwaves from the axions were produced, they would be amplified and recorded. Based on their frequency, the axion mass would then be known and its contribution to dark matter calculated. If the axion turned out to make up a considerable chunk, it would be a glorious day for particle astrophysics.

Another axion-hunting experiment, called PVLAS (Polarization of the Vacuum with Lasers), raised hope in 2006 when team members reported an unexpected twist in the polarization angle (direction of the electric and magnetic fields) of a beam of light as it passed through a strong magnetic field. They interpreted this as a possible sign that an axion may have interacted with the photon. After upgrading their equipment, however, and collecting more data they realized that their original findings were just an "instrumental artifact" and had no physical meaning.

PVLAS, ADMX, and several other groups are still competing to be the first to capture signs of the wispy particle. The soap opera—or should we say "detergent opera"—of the axion carries on and won't be over until the photon sings.

Gravity on Steroids

Some physicists are betting that WIMPs, axions, and other forms of cold dark matter won't be found at all. They are placing wagers on alternative hypotheses that modify Newton's law of gravity and Einstein's general theory of relativity under certain circumstances. By bolstering gravity in the very places where stars and galaxies are getting unseen extra tugs, they hope to avoid the need for invisible material altogether.

Under ordinary circumstances, gravity has long been known to obey an inverse-squared law—meaning that it gets weaker with the square of the distance between two objects. That would mean that the gravitational force between bodies that are very far away from each other would be extremely weak. But what if somehow gravity gets a boost under certain conditions that allow it to maintain some of its strength even for great distances? It is like a runner quaffing an energy drink halfway into a marathon and sprinting when she should be languishing. Is that legal? It depends on the rules of the game.

In 1983, Israeli physicist Mordechai Milgrom noted that stars in the outer reaches of galaxies accelerate at a much lower rate than more central stars and also much less than outer planets in the solar system. (The meaning of "accelerate" in this case doesn't mean speeding up; rather, it means turning. Stars farther from the center revolve more gradually.) He wondered if the reduced acceleration could play a part in tweaking the law of gravity. His goal was to develop an alternative explanation of the "flat rotation curves" (level plateaus when plotting speed vs. distance for peripheral stars) found by Vera Rubin that kicked off the modern dark matter quest.

Milgrom came up with an amended form of Newton's laws that he called MOND (Modified Newtonian Dynamics). In his theory, if acceleration is extremely low, the force of gravity becomes amplified by an extra factor. He calculated that the modification would keep stars moving at constant velocities, no matter how far they are away from the galactic hub. The modification thereby reproduced the "flat rotation curves" without requiring dark matter.

Critics of MOND soon noted that it violates long-held principles of physics, such as the law of momentum (mass times velocity) conservation. Momentum conservation describes how all known objects interact—in collisions, for instance. It tells us how fireworks burst, pool balls hit, and bumper cars slam. Given the law's multicentury track record for solid predictions, few physicists would consider nixing it without substantial evidence to the contrary. MOND also does not take into account the relativistic theories developed by Einstein and therefore cannot be a complete description of gravity.

To help address these concerns, in 2004 Jacob Bekenstein proposed a theory called TeVeS (tensor-vector-scalar) gravity. He mixed different kinds of mathematical objects with distinct transformational properties—tensors, vectors, and scalars—to create a variation of general relativity that produces flat rotation curves for stars in the outer reaches of galaxies. Bekenstein demonstrated that his theory reproduced many of Milgrom's results without carrying the baggage of defying sacrosanct physical principles. For example, unlike MOND, TeVeS preserves the law of momentum conservation.

Although some might find the TeVeS concoction a tasty alternative to particles named after a laundry product, the proof of the pudding is in the testing. If standard general relativity is absolutely correct in predicting how celestial objects move, then any modification of the theory must be wrong. Einstein ran up against that problem several times when he tried to tweak general relativity to unify it with electromagnetism. Whenever he attempted to adjust its framework he ran into trouble and had to retract his modifications. Like the Sistine Chapel, perfection needs no added flourishes. Nowadays, tests of general relativity have become far more precise than in Einstein's day. Every astronomical observation and physical experiment conducted so far has shown that Einstein's 1915 masterpiece requires no extra strokes.

In March 2010, a team of researchers from Princeton, Zurich, and Berkeley announced the results of a brutally rigorous fitness test for a ninety-five-year-old theory. Only the Jack LaLanne of physical models could meet such demanding standards. (The fitness guru, who has since passed away, was similarly ninety-five at the time.) Using data from the Sloan Digital Sky Survey, representing the positions and redshifts of more than seventy thousand galaxies, they combined several different measures to come up with a single score for general relativity's prognostic prowess. These measures included the bending of light by galaxies (shown through gravitational lensing), the clustering of galaxies, and the motion of galaxies due to gravitational tugs by other galaxies. The bending of light provides an indication of the warping of space due to the mass within it. According to general relativity, such warping should affect how galaxies move in predictable ways. By comparing the light-bending to the galaxy-shifting, the researchers established a benchmark for general relativity to match. Einstein's nonagenarian theory passed the fitness test magnificently with nary a cough or a wheeze! By contrast, the six-year-old TeVeS missed the mark. Before writing off any alternatives to general relativity, fledgling models sometimes develop into viable contenders. Until dark matter is identified, expect more competitors to join the fray.

9

What Is Tugging on Galaxies?

The Mysteries of Dark Flow and the Great Attractor

> The clusters show a small but measurable velocity that is independent of the universe's expansion and does not change as distances increase. . . . We never expected to find anything like this. . . . The distribution of matter in the observed universe cannot account for this motion. . . . Because the dark flow already extends so far, it likely extends across the visible universe.

—ALEXANDER KASHLINSKY, GODDARD SPACE FLIGHT CENTER PRESS RELEASE, SEPTEMBER 2008

Modern cosmology assumes that space, on its largest scale, is as smooth as porridge. Although the cosmos is full of structure—planetary systems, galaxies, clusters (groups of galaxies), and superclusters (groups of clusters)—these are like raisins, nuts, banana slices, and other chunks randomly mixed throughout the cosmic farina. Just as each ladleful of a well-mixed breakfast crock should have roughly the same ingredients, each angular slice of the nighttime sky should have about the same composition. Consequently, theories of how the entire universe behaves presume that, on average, every direction looks pretty much the same. Such overall isotropy allows cosmologists to use the relatively simple Friedmann models (including a cosmological constant) to model cosmic behavior. In these models, space expands equally in all directions.

Considering these expectations of large-scale uniformity, when three-dimensional galaxy surveys began in the late 1970s and began to gather results in the 1980s, astronomers were astonished to discover the richness and complexity of structure in space—well beyond that of clusters and superclusters. Redshift surveys by Margaret Geller and John Huchra of the Harvard-Smithsonian Center for Astrophysics, along with their research colleagues and students, discovered a jaw-dropping array of features across the sky. Within the slice of space they examined, representing about 18,000 galaxies, they found sinuous filaments and stretched-out sheets. Giant bubbles sprinkled with galaxies surrounded strangely empty voids. The cosmos seemed spongy rather than smooth.

The best-known feature they identified (and announced in 1989) was an incredibly long cordon of galaxies dubbed the "Great Wall." Spanning 600 million light-years across, it represented the largest structure found in space until that point—and still one of the most sizable known today. Only the currents of pure chance coupled with the unifying force of gravity could explain how such a formidable grouping assembled itself.

While the discovery of gargantuan formations does not negate the uniformity of space on its largest scale, it shows that on a wide range of scales—up to hundreds of millions of light-years—it is astonishingly diverse. Such diversity makes itself known in the flow of galaxies. Uneven clumping means unequal gravitational pull, producing imbalanced tugs on galaxies. This can cause galaxies to sway from the purely outward motion expected due to space's expansion. Astronomers call the extra movements of galaxies (in addition to Hubble expansion) "peculiar velocities."

Rather than simply attributing cosmic density extremes to chance, researchers are increasingly looking for possible underlying causes. It is like discovering a world in which most people reside in an enormous megalopolis and almost nobody lives in certain vast, deserted regions. Such a discrepancy could be sheer chance—or the result of environmental factors. Could there be "environmental causes" that have segregated the universe into unevenly populated regions?

If the observable universe were all that were out there, it would be odd to consider "environmental causes." However, increasingly astronomers have begun to reckon with the possibility that there is much more to the universe than we can observe. Even if we cannot directly observe other parts of the multiverse, perhaps they are affecting us through their hidden connections.

An analogy for such a situation is a classroom that has a window positioned so that while some of the students can see out of it, the teacher cannot. If an ice cream truck parks outside the window, the teacher might notice a general craning of necks and turning of heads in that direction and wonder what is going on. Even if she can't witness what is happening outside, the disruption of the order of the classroom would offer her a clue as to the influence from beyond the room. Similarly, the hidden parts of the multiverse, banished beyond observation, still could make their presence felt through their gravitational attraction and other effects.

Beyond the Zone of Avoidance

In the early 1990s, astronomers used the COBE measurements of the cosmic microwave background to determine the peculiar velocity of the Local Group of galaxies. The Local Group includes the Milky Way, Andromeda, and a number of smaller galaxies. Because of their mutual gravitational attraction, these galactic mates tend to travel together through space. Researchers estimated that the Local Group has a peculiar velocity of about 400 miles per second and is aimed roughly in the direction of the Virgo cluster (a larger grouping of galaxies).

Astronomers are baffled as to what cosmic dynamo is hauling the Local Group through the depths of space. At first the Virgo cluster seemed a likely culprit. However, calculations revealed that it wasn't massive enough or in the right position to offer enough of a yank. Something else must be offering a gravitational boost. One possibility is undetected dark matter. Perhaps a massive distribution of unseen material is tugging on our galaxy and its neighbors.

The peculiar velocity of the Local Group is but a part of an even larger flow. In the 1980s a Caltech collaboration discovered a colossal galaxy magnet called the Great Attractor, a patch of space toward which the Milky Way and tens of thousands of other galaxies are racing at an incredible speed of 14 million miles per hour. What precisely is driving this galactic steeplechase remains unclear.

Analysis of what actually lies in the Great Attractor has been stymied by its "cheap seat" location—beyond the dusty disk of the Milky Way and out of clear view. Because of their difficulty in seeing beyond it, astronomers call that sector of the sky the Zone of Avoidance. Nevertheless, progress has been made in mapping that part of the celestial dome.

In the late 1980s, while looking for the source of the Great Attractor, several teams of astronomers identified one of the largest galactic groupings in space, called the Shapley supercluster. Indian astrophysicist Somak Raychaudhury, one of the first to examine the Shapley supercluster, determined that it was too distant to serve as the primary cause of the Great Attractor. However, in the mid-1990s NASA researchers managed to peer through the veil of the Zone of Avoidance by use of the ROSAT X-ray satellite and discover the massive size of the Abell 3627 cluster. Although that cluster had previously been observed, its huge extent had been unknown because of the Milky Way dust obscuring the view. Current thinking is that the Abell 3627 cluster lies at the heart of another large supercluster with strong enough gravitational pull to cause the mammoth flow of galaxies associated with the Great Attractor.

To map out what lies beyond the Zone of Avoidance, a collaboration of radio astronomers led by Patricia A. (Trish) Henning of the University of New Mexico have aimed the Arecibo Radio Telescope at that part of the sky and recorded the signals of occluded galaxies. Nestled in the lush rain forest of northern Puerto Rico, the Arecibo Radio Telescope is the world's largest and most powerful such instrument with a single antenna. Therefore it is ideal for a radio survey of remote signals from masked galaxies. While the dense center of the Milky Way cloaks most of the optical light from these galaxies, radio signals from their swirling hydrogen gas can get through. Beginning in 2008, Henning's team has been profiling how these hidden galaxies are distributed, with the hope of understanding the source of the Great Attractor and other unexplained gravitational lures.

The Great Void in Eridanus

Since the 1980s and 1990s, voids, walls, filaments, attractors, and other anomalies have become part of the essential lexicon of astronomical description. One critical tool in finding these structures has been the biennial release of cosmic microwave background data from the WMAP satellite. Astronomers have perused each report, looking for evidence of unusual patterns of relatively colder and "warmer" spots. (The background is only 2.728 degrees Kelvin above absolute zero, so "warmer" means only in the millionths of a degree higher than average). The colder spots are matched to the places they correspond to in the sky, which are examined for underdense regions.

In 2007, radio astronomer Lawrence Rudnick unexpectedly discovered one of the largest voids ever found by aiming a receiver in the direction of a WMAP cold spot. As he described the experience, one day he had some extra time on his hands, decided to check out the WMAP data, and thought it might be interesting to search the region of the constellation Eridanus, where a sizable, colder-than-average patch of the CMB is located. He and his team were astonished when, after scanning a sky sector almost 1 billion light-years across, they found virtually no radio sources. Radio signals tend to correlate with populous parts of space. Therefore, the lack of radio emissions seemed to sketch out an incredibly empty region spanning a large part of the sky. The Great Void in Eridanus seemed truly immense!

What could have poked such a vast hole in the celestial firmament? One speculative theory, proposed by physicists Laura Mersini-Houghton of the University of North Carolina at Chapel Hill and Richard Holman of Carnegie-Mellon University shortly after Rudnick's discovery, is that the colossal void is the imprint of a primordial quantum interaction with another region of the cosmos that has since become so widely separated from ours that we cannot observe it. The primitive vacuum, they asserted, included entangled quantum states that exerted a long-range influence on each other. Entanglement occurs when two quantum states are linked to each other in a kind of symbiotic relationship; if one changes, the other must alter as well, even if it corresponds to an object physically remote from the first object. Einstein called this "spooky action at a distance." Mersini-Houghton and Holman postulated that such entanglement served as a kind of conduit of vacuum energy from a neighboring patch of the cosmos to our own, causing negative pressure that inhibited galaxies from seeding in the region that became the void. It is like finding that a stack of papers refuses to sit on a desk—constantly blowing away and leaving it bare—and then discovering a hole in the wall behind it through which winds are gusting from a neighboring dwelling's capacious air conditioner.

The idea of interactions with an alternate universe has long been explored in fiction. For example, in Isaac Asimov's 1973 novel *The Gods Themselves*, a civilization of the future taps into an energy source leaking into our universe from another realm called the parauniverse. That trans-cosmic exchange proves disruptive for both realities. Could hidden sectors of the cosmos be funneling energy or information into ours, creating gigantic forbidden zones in the process? Or could there be a simpler explanation for the Saharas of space?

Voids of such immensity were so unexpected that researchers scrambled to check and recheck the data. In 2008, Kendrick Smith of the University of Cambridge and Dragan Huterer of the University of Michigan called Rudnick's analysis into question and raised doubt about the existence of the colossal empty region. They alleged that Rudnick's group undercounted the number of galaxies in that part of space by excluding weaker signals. When Smith and Huterer considered all candidate galaxies in that sector, including dimmer ones, the count seemed close to the expected number for a region of that size. Could that grand cosmic desert have been a mirage?

Astronomers continue to debate the reality and extent of the Eridanus void. No one doubts, however, the existence of the large WMAP cold spot to which it corresponds. Whether it was quantum entanglement or another mechanism that wiped the slate clean is a matter of ongoing research.

Meanwhile, an even larger void was discovered in 2009 during sky scans by the Six Degree Field Galaxy Survey (6dFGS), a project based at the 4-foot-diameter UK Schmidt Telescope in Australia. Led by Heath

Jones of the Anglo-Australian Observatory, the discovery team included void pioneer Huchra and dozens of other researchers from around the world. (Sadly, Huchra passed away in October 2010.) The survey covered 41 percent of the sky, including some 110,000 galaxies, probing as far as 2 billion light-years from Earth. In an incredible result, they identified a void in a section of the sky above the Southern Hemisphere that is 3.5 billion light-years across. The extraordinary size of that hole in space makes it extremely hard to understand how it developed during the 13.75 billion years since the Big Bang. The group also contributed a more detailed map of the Shapley supercluster—its immense, overpopulated region representing the opposite extreme.

Reports from the 6dFGS and other recent surveys have defied expectations of a uniform universe. Increasingly, researchers are grappling with the possibility that unseen forces beyond the edge of the observable universe have influenced the structures within it, shattering its uniformity through hidden tugs.

Dark Flow to Reaches Beyond

A startling astronomical discovery made in 2008 may represent the smoking gun for the existence of regions of the universe beyond the cosmic horizon. A research team led by Alexander Kashlinsky of NASA's Goddard Space Flight Center in Greenbelt, Maryland, applied a special statistical analysis of microwave radiation left over from the Big Bang, collected by the WMAP satellite, to examine how clusters of galaxies move through space. The CMB data revealed an astonishingly vast flow of clusters along a single direction, unexplained by objects within the observable universe.

The cosmic microwave background has proven to be an extraordinarily useful tool for understanding structure in space. Not only does it offer vital insight into the way the universe was more than 13 billion years ago, it also serves as an instrument for discerning how clusters of galaxies move through space right now. That is because the CMB is far from a static relic of the era of recombination (when neutral atoms formed and radiation first became free of matter). Rather, as it expands and cools down with the growth of the universe, it sometimes comes into contact with hot gases—particularly the scalding froth bathing the galaxies in clusters—heated within each cluster's gravitational tub. When photons from the CMB interact with electrons from these gases they scatter like falling raindrops hitting blasts of steam rising from a vent. This interaction boosts the photons' energies—a phenomenon first studied in the 1970s by Soviet physicists Rashid Sunyaev and Yakov Zel'dovich and called the Sunyaev-Zel'dovich Effect (SZE).

There are two components to the SZE, the thermal and the kinematic parts. The thermal SZE, first detected in the 1980s, involves the photons from the cosmic background gaining energy from the hot gas. Much harder to measure, the kinematic SZE constitutes changes in the photons' wavelengths (distance between peaks) due to the motion of the gas relative to the cosmic background. Since the cosmic background radiation is expanding with space, if the large-scale movements of clusters are due only to the overall growth of the universe, such an effect should not appear. On the other hand, if the clusters have motion in addition to the cosmic expansion, the kinematic SZE should pick up those movements through wavelength shifts.

Situated near the automotive Beltway that rings the District of Columbia, NASA's Goddard Space Flight Center should be used to large-scale movement—at least in theory. While the stop-and-go traffic around that loop defies all attempts at prognosis, Kashlinsky's research group has become adept at mapping out far grander movements in space that take place much more regularly.

When there is movement on the Beltway, we know what drives it: internal-combustion engines powering vehicles taking residents to reasonable destinations such as work, home, or Maryland beaches. The astronomical motions Kashlinsky's team are looking at—the parade of clusters through space—are being driven by an unknown agent and heading toward in an inexplicable direction. Therefore, while its flow is smoother than Beltway traffic, its motivation is as unclear as that of some of the federal agencies within that ring road.

In 2008, Kashlinsky's group conducted a study of 782 clusters that emit radiation in the X-ray range. The catalog they used was one of the largest compilations of X-ray cluster data ever assembled. The researchers applied the kinematic SZE method to the WMAP three-year data and looked for temperature shifts in the cosmic microwave background in the regions corresponding to the clusters. Kashlinsky's team detected these in droves—recording significant shifts in the CMB throughout a region more than 1 billion light-years across. These shifts appeared only in the vicinity of clusters, signifying that they were likely due to the clusters' motion through space relative to the Hubble expansion.

To the researchers' astonishment, they found that hundreds of clusters are being drawn toward a certain part of the sky at 2.2 million miles per hour. It is as if a plug had been pulled out of that region and the clusters have been streaming toward it. Nothing known in astronomy could be causing such a colossal movement. As far out as they looked, this unidirectional bulk flow showed no signs of tapering off. Nothing known in astronomy could be causing such an extraordinarily vast movement. The researchers speculated that it could be a tug from matter beyond the observable universe. In homage to the mysteries of dark matter and dark energy, Kashlinsky dubbed this unexplained motion "dark flow."

A follow-up study in 2010, based on updated WMAP data, revealed an even more startling picture. Examining fourteen hundred clusters of galaxies, Kashlinsky and his team found that dark flow extends at least twice as far across the sky as they had originally surmised. They determined its extent to be more than 2.5 billion light-years out from Earth. Kashlinsky conjectured that dark flow reached even farther out—to the edge of the observable universe. What could be tugging on so many clusters of galaxies?

Canadian astrophysicist Mike Hudson of the University of Waterloo, along with his collaborators, has independently detected other large-scale movements of galaxies that seem inconsistent with the standard model of cosmology. His team has measured the peculiar velocities for a substantial region of the sky—from our galactic neighborhood out to 400 million light-years—and found an overall cosmic flow that defies explanation. Interestingly, the large-scale movement of galaxies found by Hudson's team is roughly in the same direction as the dark flow detected by Kashlinsky's group. Hudson has pointed out that while there is a small chance that the isotropic, homogeneous Big Bang picture could accommodate such large flows, statistics indicate a greater likelihood that the standard model will need to be revised.

As Hudson remarked, "It's as if, in addition to the expansion, our 'neighbourhood' in the universe has an extra kick in a certain direction.

We expected the expansion to become more uniform on increasingly larger scales, but that's not what we found."[1]

Kashlinsky and his colleagues have offered an intriguing hypothesis for the cause of dark flow, based on the inflationary universe model. As they wrote:

> An interesting, if exotic, explanation for such a "dark flow" would come naturally within certain inflationary models. In general, within these models, the observable Universe represents part of a homogeneous inflated region embedded in an inhomogeneous space-time. On scales much larger than the Hubble radius, pre-inflationary remnants can induce ... CMB anisotropies. ... [Such a situation] would lead to a uniform flow.[2]

In other words, inflation, the ultrarapid stretching of the universe shortly after its birth, would whisk inhomogeneities (irregularities from place to place) beyond the horizon of the observable universe—the region within which we could ever detect its light. The inhomogeneities that were close to our part of space before inflation would now be outside the observable zone and couldn't be detected directly. What we did observe, therefore, would be relatively homogeneous and isotropic (appearing the same in all directions). However, despite the uniformity, there still could be distant gravitational influences from the irregular reaches beyond, leading to dark flow. Massive heaps on the sidelines, pushed aside by inflation, could have enough gravitational pull to influence clusters in our space in an irregular way, causing them to move in a particular direction. It is like plowing a snowy street in winter to make it completely flat and finding that the meltwater from the piled snow keeps seeping in, refreezing and leaving tracks—and causing drivers to skid. Could clusters in our universe be "skidding" toward the mounds left by inflation?

Dark flow has not yet acquired the pedigree assigned to dark matter and dark energy as the leading mysteries of cosmology. However, it points to an important issue with inflationary cosmology that it might not be able to sweep all the relics of the stage before inflation under the carpet. If dark flow does provide proof of a multiverse—other universes beyond ours—it would represent an extraordinary find indeed. Along with dark matter, dark energy, cold spots, and giant voids, dark flow provides even further evidence that the standard model of cosmology—successful as it is in describing the age and evolution of the universe—is far from complete. Revolutionary change is in the air; where it will lead remains unknown.

Intruder Alert

For generations, cosmology has assumed that all parts of space are essentially the same, give or take the particular distributions of galaxies. Newfound anomalies, however, have challenged the field to rethink the Copernican principle and the notion of overall isotropy. Could there be—as dark flow seems to suggest—a preferred direction in space? Might it be the case—as the presence of enormous voids, formidable cold spots in the CMB, gargantuan superclusters, and colossal walls of galaxies seem to indicate—that the universe is really inhomogeneous? General relativity works simplest with isotropic, homogeneous cosmology, but with each new discovery of unevenness, researchers have more reason to reevaluate these assumptions.

The possibility of favored directions in space naturally suggests that forces from regions beyond observation are creating such arrows. Interactions with spaces outside our own bubble universe fundamentally contradict the reason for postulating an inflationary era in the first place. Inflation was supposed to smooth out the observable universe to the extent that there would be no record of times before and domains beyond. One would need incredible fine-tuning for inflation almost to level the playing field but to leave some anisotropies.

Indeed, inflation was meant to allow us to forget about other possible universes, with their ghastly irregularities, and focus on our finely honed diamond. But once Linde demonstrated how simple it is for inflationary domains to be seeded, he and many other cosmologists began exploring the possibility that inflation is eternally taking place, spawning more and more universes. The possibility that we live in a multiverse has become a hot topic of discussion in theoretical cosmology, despite inclinations to be cautious. And if we do live in a cosmic hall of mirrors, could we be occasionally catching glimpses of other chambers?

As Carl Sagan famously remarked, "extraordinary claims require extraordinary evidence."[3] Before hitching modern cosmology to the multiverse bandwagon, scientists must be extremely cautious in interpreting any data that suggest outside influences. They need to rule out any of a wide range of more mundane explanations and the ever-present possibility of statistical flukes.

In general, the study of the universe has entered an exciting but perilous new phase due to the increasingly greater use of statistics in analyzing increasingly complex sets of astronomical data. While opportunities are arising for mind-blowing discoveries, we must take heed not to overinterpret chance combinations of factors and start seeing phantom phenomena where none actually exist. Fortunately, the new generation of cosmologists is, by and large, learning how to be cautious in its claims.

10

What Is the "Axis of Evil"?

Investigating Strange Features of the Cosmic Background

There have been a number of disturbing claims of evidence
for a preferred direction in the Universe. . . . These claims have
potentially very damaging implications for the standard
model of cosmology.

—KATE LAND AND JOÃO MAGUEIJO, *THE AXIS OF EVIL*

The WMAP satellite and other probes of the cosmic microwave background have been as revolutionary for cosmology as global positioning systems (GPS) have been for navigation. Like GPS, WMAP has provided researchers with a detailed map that can be used to locate features and to scan for patterns. Just as detectives have employed GPS trackers to sleuth the driving habits of alleged criminals, hoping to find out how they operate, astronomers have delved into WMAP data seeking greater understanding of cosmological mechanisms. For example, WMAP data could offer vital clues as to whether the observable universe is part of a vast multiverse.

The key for researchers is figuring out which patterns are merely coincidence and which tell us something meaningful about the laws of the universe. We already know that the observable universe is generally the same in every direction, but there are clumps and empty parts when you look more closely. When are those clusters and voids just random chance, and when are they products of hidden influences?

WMAP data involve extremely subtle temperature variations of less than 0.0002 degree Kelvin in a frigid bath of relic radiation spanning the sky. Starting in 2006, detailed reports of the data—generally of higher and higher resolution—were released every two years. The longer the sky was tracked, the more data were collected and the more precise they could be. To help distinguish the relic radiation from galactic radio emissions, the satellite was tuned to five different frequency bands—something like five different radio channels. For the highest-frequency channel, 90 gigahertz, the satellite could distinguish temperature differences in points of the sky fewer than 0.25 angular degree apart. For the lower-frequency channels the angular resolution was not quite as precise. All in all, the temperature fluctuation maps of the sky have included millions of data points. To interpret this information has required powerful statistical techniques capable of sorting through haystacks of data, looking for any needles of correlation pointing to unexpected phenomena.

Ever since the first comprehensive WMAP results were announced, astronomers have pored through the data, looking for anomalies. Searching for colder than average regions corresponding to voids such as the Great Void in Eridanus and looking for the telltale energy boosts (in the kinematic Sunyaev-Zel'dovich Effect) that point to the dark flow of clusters are but two of the ways cosmologists have put the WMAP data to good use. Experimental scientists are generally a cautious lot, inclined to announce results only after they have put them through their paces by running one statistical analysis after another. Any purported finding carries with it a chance of turning out to be a fluke. The case of Hawking's initials illustrates that point well.

Hawking's Monogram in the Sky

A curious lesson about statistics in cosmology surfaced at about the time of Stephen Hawking's retirement as Lucasian chair of mathematics at the University of Cambridge. Like a monogram stitched into in a hand-sewn retirement present, his initials appeared in the WMAP data. Nature seemed to be honoring an illustrious career that spanned many decades of groundbreaking research despite profound physical limitations. Hawking had held that prestigious appointment—a professorship once held by Sir Isaac Newton—from 1979 until 2009.

Shortly after he was appointed to that position, Hawking signed his name by hand for the last time. His disability, amyotrophic lateral sclerosis (ALS), also known as Lou Gehrig's disease, had progressively robbed him of his motor functioning, leaving him confined to a wheelchair and increasingly unable to communicate (and eventually needing a computer system for this purpose). With weakened grip and unsteady hand, Hawking still managed to sign the "Admission to Office" book for his prestigious new role. As he recalled,

> They have a big book which every university teaching officer is supposed to sign. After I had been Lucasian Professor for about a year, they realized I had never signed. So they brought the book to my office and I signed with some difficulty. That was the last time I signed my name.[1]

Hawking's three-decade-long Lucasian professorship corresponded to a period of extraordinary change in cosmology. In October 2009, Hawking retired from that position. He was soon to turn sixty-seven, which was the mandatory retirement age. Although no longer Lucasian professor, he maintained a research position at Cambridge and assumed a new position at the Perimeter Institute in Canada, where the Stephen Hawking Centre was established in his honor.

Only a few months after Hawking stepped down as Lucasian professor and shortly after his sixty-seventh birthday, his "signature" turned up rather unexpectedly in a record far more ancient than the Cambridge "Admission to Office" book. More precisely, it was his initials—perfectly aligned, evenly spaced, and seemingly composed in similar font sizes and styles—in a portrait of what the cosmos looked like more than 13 billion years ago. Unless he had secretly mastered the art of time travel and left the imprint as proof, how could that be?

The picture of the sky's ancient light, with Hawking's initials clearly seen, was tucked away into a rather serious-sounding research article, "Seven Year Wilkinson Microwave Anisotropy Probe (WMAP) Observations: Are There Cosmic Microwave Background Anomalies?," which was posted on an online archive in January 2010 and later published in the prestigious *Astrophysical Journal Letters*. As the researchers reported,

> Shortly after the WMAP sky maps became available, one of the authors [Lyman Page] noted that the initials of Stephen Hawking appear in the temperature map. . . . Both the "S" and "H" are beautifully vertical in Galactic coordinates, spaced consistently. . . . We pose the question, what is the probability of this occurrence? It is certainly infinitesimal.[2]

In addition to Hawking's initials in the sky, the article addressed a variety of curious findings from the WMAP probe. Other seeming anomalies included prominent, relatively large cold spots and unexpected alignments of various angular components.

Although the authors of the paper stressed the improbability of Hawking's initials appearing so clearly in the sky map, they emphasized that with any enormous amount of data, extremely unlikely events are bound to occur. They highlighted the curiously "autographed" results mainly to make that important point. As they wrote,

> [The appearance of the initials] is much less likely than several claimed cosmological anomalies. Yet, we do not take this anomaly seriously because it is silly. The Stephen Hawking initials outline the problem with *a posteriori* statistics. By looking at a rich data set in multiple different ways *unlikely events are expected*. The search for statistical oddities must be viewed differently from tests of predetermined hypotheses.[3]

In other words, it is important to distinguish things you are looking for from things that unexpectedly turn up. If you are searching for something and it shows up in carefully analyzed data, there is a much greater chance of it being an important finding than if you come across something you never anticipated. While the latter could be genuine science, it also could be a random oddity.

Cosmic sleuths often pore through astronomical information looking for anomalies, with the aspiration of identifying some hitherto unknown feature of the universe. If they are lucky, they might discover an incredible new aspect of the cosmos that could revolutionize science. The chance finding might lead to a pattern that could overthrow existing cosmological models. If not, the researchers might move on to another curiosity. While no one seriously considered locating Hawking's initials a genuine scientific breakthrough, there are other anomalies for which scientists are unsure of their significance. If these proved important, theorists would need to rethink existing theories of the universe.

The Axis of Evil

Finding large cold spots and identifying the dark flow of clusters have been two of the notable outcomes of the very thorough statistical searches for unusual sections of the CMB. Additionally, there have been even subtler curiosities that have continued to defy understanding. One of these is the "axis of evil"—a strange alignment of multipoles first identified by Kate Land and João Magueijo.

Multipoles are types of three-dimensional ripples equivalent to standing wave (vibrating in place) patterns in two dimensions. If you open a piano and pluck one of its strings, you might create a single peak, two peaks, or some other number. These are called the harmonics. Playing a note on a piano generally produces some combination of harmonics. If you analyze a note, you can break it down into what proportion of each harmonic forms the total mix.

Now hit a balloon with a mallet to create ripples on its surface. These would constitute various wobbling portions of the balloon's exterior, vibrating in and out like a frog's throat. The basic set of these, from the simplest to higher orders, are called the spherical multipoles. Any type of surface irregularity can be expressed as a combination of multipoles.

The most elementary multipole is the dipole, consisting of a shift in a single direction, like taking a balloon and compressing one side while the other expands. In the analysis of the CMB, that was the first multipole to be found—attributed to Earth's motion through space. Following dipoles, the next order of multipoles are quadrupoles, octopoles, and so forth. Higher-order multipoles represent subtler types of deviations from a purely spherical form—finer and finer patterns of bumps, like hills and valleys on the surface of Earth. Given WMAP's ability to collect precise data about minute temperature fluctuations in different parts of the sky, astronomers have found that a breakdown of the information into multipoles—a classification called a power spectrum—is an ideal way to analyze these wrinkles. The power spectrum helps astronomers understand how different attributes of the cosmos, such as its curvature, early history, and composition, come into play. Matching it against the Lambda Cold Dark Matter model has offered evidence that the standard model of the cosmos is a good fit and that researchers have generally been on the right track. However, certain subtle aspects of the power spectrum remain inexplicable—such as the curious axis of evil alignment.

Land was a PhD student working with Magueijo at Imperial College, University of London, when they decided to tackle the question of possible anomalies in the cosmic microwave background. Born in Sussex, England, she had become interested in astronomy at a young age. As a child, she sometimes spent sleepless nights pondering cosmic questions such as "Where is the edge of the universe?" and "What is empty space made of?"[4]

The Portuguese-born Magueijo is well known in the cosmology community for his maverick ideas. In 1998, he and Andreas Albrecht put forth a theory of the universe called the Varying Speed of Light (VSL) model, which set forth to solve the horizon, flatness, and cosmological constant problems without the need for an inflationary epoch. It suggested nothing less than replacing Einstein's special theory of relativity with an alternative approach that allows the speed of light to change over time. Without dismissing the possibility altogether, South African mathematician George Ellis has pointed out that tinkering with the speed of light would greatly affect many of the laws of physics and would have numerous potentially observable consequences.[5]

In looking at CMB anomalies, Land and Magueijo set out along a more technical path, based on statistical analyses rather than new theories. Nevertheless, it was a road that could lead to revolutionary conclusions if they turned up any markers pointing to any discrepancies from the standard interpretation. The earlier COBE data had indicated an unexpected alignment between the quadrupole and octopole moments; researchers weren't sure what it meant. A 2003 paper by Max Tegmark, Angelica de Oliveira-Costa, and Andrew J. S. Hamilton similarly found an unexplained alignment in early WMAP results.[6] Land realized that an analysis of alignments in the WMAP data could potentially reveal something new about the cosmos.

"If there was anything 'odd' on the largest scales of the Universe," she said, "then this would the first place you'd expect to find any signature."[7]

Amazingly, they found more than they had bargained for. Not only were the first two multipole modes aligned along a particular axis, but also the next two higher ones were lined up with them. For some reason, patterns of cold and warm spots were forming a line along a particular direction of space. There was no obvious reason for such an alignment, but the result was intriguing.

To describe the peculiar alignment, Magueijo came up with the name "axis of evil" as a humorous reference to the news of the day. As Land recalled,

> This research was taking place during the Iraq invasion and Bush was all over the news at the time talking about the Axis of Evil. We started calling the anomalous alignment the "axis of evil" as a bit of a joke really . . . then it became the working title of our paper. And by the time we submitted the paper it had stuck with us (and colleagues) so much that we didn't change it! But the journal did change the title.[8]

Thanks, perhaps, to the evocative name for the alignment, the paper received considerable notice. As Land related, "People laughed at the 'axis of evil' title. I think the title really helped advertise the work—for better or worse, a catchy name really helps someone remember the topic and you get a lot of citations!"[9]

Reaction from the researchers associated with the WMAP project was more guarded. They were naturally very protective about the data and the data's analysis: they wished to avoid any errors associated with the probe, how it collected and transmitted information, and how that information was interpreted. As in the example of Hawking's initials seen in the sky, the WMAP astronomers often caution against overinterpretation of chance patterns.

To ensure that what they had found was significant, Land and Magueijo decided to conduct a new analysis based on a method, called Bayesian statistics, that judges the significance of patterns on the basis of prior expectations as well as the data itself. It penalizes for trying to include too many factors when trying to fit the data if they aren't well justified. By comparison, the first method they used, called frequentist statistics, compares the likelihood of what is found to the set of all possibilities and does not penalize for including extra factors. Bayesian statistics can be used as a hedge against overfitting data—meaning developing a formula that perfectly matches every piece of evidence but does little to predict the future. It cautions that the more exact your prescription for what is going on, without sufficient reason to add such precise constraints, the less likely it is that it will hold for other cases.

For example, suppose that one day you see kids, all wearing uniforms, leaving a school. The next day you observe the same thing. You note that in almost every case you've seen or read about where kids wear uniforms, there is a mandate to do so. Therefore you conclude that the school has a dress code. Based on your prior experience, Bayesian analysis encourages you to make that simple conclusion.

However, let's say you are standing near another school every day of the week during dismissal time. On a Tuesday you see a girl wearing red shoes, on a Wednesday you see a boy with a blue shirt, and on a Thursday you see a boy chasing another boy who has a green corduroy jacket. You try to take into account each of these parameters and concoct an elaborate theory relating days, colors, clothes, and behavior. You develop a hypothesis that if a girl with red shoes and a boy with a shirt other than red leave the school at 3:00 P.M. on consecutive days, there must be at least someone wearing corduroy, of yet another color, running out the door at 3:05 P.M. on the third day. That unjustifiably close level of matching is called overfitting the data. Based on your prior experiences, you would have little basis for adding so many specific factors to your prediction. Therefore the Bayesian approach would advise that the amount of detail you put into the model would make it less likely to be accurate.

Applied to the "axis of evil" hypothesis, the Bayesian analysis dampened expectations that the alignment represented an actual physical phenomenon rather than chance. That is because the researchers added some extra parameters to fit the data that did not improve the likelihood. Therefore, while in their follow-up paper, called "The Axis of Evil Revisited," the researchers confirmed that the alignment does indeed exist, they could not point to statistics that bolstered its likelihood of being a real effect. As Land remarked,

> If I had to bet my life on it, I would say nothing has caused the alignment. It "exists" in the data, but it is just by chance. And our first paper perhaps overstated the significance of the feature. If it is caused by something real, then I would guess some effect of large scale structure around us is warping the observations of the CMB as the light travels towards us.[10]

Analysis continues on whether there is a physical reason for the "axis of evil." Adding to the mystery is a 2007 finding by Michael Longo of the University of Michigan that the rotational axes of spiral galaxies tend to line up in the general direction of the "axis of evil." Based on Sloan Digital Sky Survey data, he determined the axes of spin of thousands of spiral galaxies and found that most of these tilted in a direction roughly corresponding with the alignment found by Land and Magueijo. Although the effect is approximate, rather than something stark and immediately noticeable, it was still improbable enough to compel further study of the "axis of evil" to see if it might be a genuine physical phenomenon.

If the "axis of evil" does turn out to be real, astronomers will be motivated to try to identify cosmic structures large enough to cause such an alignment. The secret could be buried in the distant past. Perhaps it is a scar left over from a more chaotic time in the early universe. Just as rocky formations jutting from a flat landscape could be remnants of ancient volcanic activity, anomalous configurations in the CMB could point to primordial cosmic turbulence. Time will tell if the "axis of evil" is a vital cosmological clue or as inconsequential as Hawking's initials in the sky.

The Multiverse Shows Its Spots

One of the latest and most promising applications for the CMB data is the search for proof of the multiverse. When Alexander Vilenkin, Andrei Linde, and others suggested in the 1980s that inflation could be eternal, continuously producing bubble universes, the concept seemed abstract and experimentally unverifiable. After all, if the other bubbles are unreachable, how could we know they are out there?

Indirectly, we might surmise that the multiverse exists. As we've discussed, the extreme unlikelihood of the cosmological constant being so small, compared to its high calculated value in most theoretical models of the vacuum, has led physicists such as Leonard Susskind to argue for a survival of the fittest among bubble universes. Berkeley researchers Lawrence J. Hall and Yasunori Nomura call the low value of the cosmological constant the "degree of unnaturalness" and note that it (along with other physical parameters with unexpected values) demonstrates that our universe must be the product of "environmental selection."[11]

A hard-nosed experimental researcher would dismiss such abstract arguments and call for more tangible proof. She would be more likely to be swayed by evidence of any discernible impact of the other bubble universes on ours. Such an impact would have occurred very early in our universe's history, before the bubbles inflated too far away to affect us. The cosmic background radiation would bear witness, perhaps, to the vestiges of such a primordial interaction.

In 2009, astrophysicist Hiranya Peiris of University College London, along with string theorist Matthew Johnson of the Perimeter Institute, decided to tackle the question of whether the CMB could be used to test the idea of bubble universes. In particular they sought the fossil imprint of collisions between bubbles—which they hoped to find as characteristic signatures in the CMB data. They applied for a grant from the Foundational Questions Institute, a group that funds unconventional, far-reaching projects, and received more than $112,000 to carry out their work.

Born in Sri Lanka, Peiris received her PhD from Princeton in 2003 under the supervision of noted cosmologist David Spergel. While at Princeton she joined the WMAP collaboration and took part in its breakthrough studies of the age and composition of the universe. She became an expert in how the WMAP data displayed the fingerprints of an era of inflation. After she met Johnson she became convinced that a search for the impact of the multiverse through a statistical analysis of the CMB data would be a worthwhile endeavor.

As Peiris remarked, "I'd heard about this 'multiverse' for years and years, and I never took it seriously because I thought it's not testable. I was just amazed by the idea that you can test for all these other universes out there—it's just mind-blowing."[12]

Along with Stephen Feeney of UCL, they developed a computer simulation of what it would look like in the CMB if two bubble universes had collided. Their simulation pointed to blotches with characteristic shapes and sizes that would reflect the conditions after such a collision. Because the collision would happen before the inflationary era—when the bubbles were sufficiently close—the blotches would be stretched out during inflation, eventually leaving noticeable imprints in the CMB.

In 2011, the team announced results from an analysis of WMAP data that employed a computer algorithm to scan for the predicted collision patterns. The researchers used Bayesian statistical methods to check for significant matches between theory and data. Curiously they found four features that could possibly represent the remnants of bubble collisions. However, they did not have enough data to declare conclusively that the patterns they discovered were statistically significant. They hope that results from the Planck satellite, currently collecting CMB data, will help support their findings. The first Planck results are expected to be released in 2013.

WMAP and other CMB probes have revealed much about the very early universe—an era that, before the satellites were launched, was clouded in mystery. The instruments offer vital snapshots from the recombination era when the background radiation was released. After that era, however, there is a substantial chunk of cosmic history that remains veiled. The "Dark Ages," the shrouded interval between the recombination era and the formation of the first stars, represents a critical new frontier of cosmology.

11

What Are the Immense Blasts of Energy from the Farthest Reaches of Space?

Gamma-Ray Bursts and the Quest for Cosmic Dragons

It does not do to leave a live dragon out of your calculations,
if you live near him.

—J.R.R. TOLKIEN, *THE HOBBIT, OR, THERE AND BACK AGAIN* (1937)

Given the immensity of the observable universe, one would expect that the farthest reaches would send us only the faintest of signals. Indeed, that is true for typical astronomical objects that are extremely remote. The exceptions are bodies of incredible power, called quasars, that stand out like flares in the blackness of space. They are not only record-holders for how much energy they churn out, but also for age. Because they are so far away, they are incredibly old, and represent some of the most ancient objects in the universe.

Radio astronomers first observed quasars in the 1960s as intense beacons of radio waves. Believing these objects to be radio emitters within our own galaxy, observers labeled them "quasi-stellar radio sources." Only gradually did the truth become apparent that quasars lie much, much farther away than the boundaries of the Milky Way. In fact, they comprise some of the most remote astronomical entities in space— revealed by their large Doppler shifts, signifying outward motion up to 80 percent of the speed of light. Their remoteness points to their ancient origins, dating as far back as 10 billion years ago. Given their immense distance and amazing brightness, astronomers reached the unmistakable conclusion that quasars are powerhouses of energy, emitting far more radiation than average galaxies. Such extraordinary torrents emerge from concentrated regions typically not much bigger than the solar system.

Astronomers speculate that the dynamos empowering quasars are supermassive black holes lying at the centers of very young galaxies. These form accretion disks of in-falling matter—whirling dervishes of material trapped in a death spiral around the gravitational sinkhole. Supermassive black holes are so compact that even one of these, millions of times heavier than the Sun (such as the object thought to lie at the center of the Milky Way), would have a diameter less than the distance between the Sun and Mercury. The supermassive black holes fueling quasars would likely be billions of times heavier than the Sun, with diameters at least twice (and possibly up to twenty times as much as) the average distance between the Sun and Neptune.

As particles swirl around the maelstrom at velocities approaching the speed of light, they emit beams of energy called synchrotron radiation, which is seen in circular particle accelerators. In synchrotron radiation, when particles revolve at fast speeds they release energetic photons. In the case of the accretion disks surrounding supermassive black holes in nascent galaxies, such radiation is extraordinarily intense, resulting in the extreme brilliance of quasars.

There are no contemporary quasars, as they represent an epoch in which galactic centers were far more turbulent. Such seething cores are known as active galactic nuclei. Some astronomers believe that quasars represent a necessary stage in the lives of galaxies.

Galaxies have changed greatly in character over more than 10 billion years of cosmic history. The first galaxies, like those now at the edge of the universe, were likely relatively small and short-lived. By crashing together, these galactic "Mini-Me's" created more substantial groupings of stars. A 3-billion-year bumper car match of colliding petite galaxies preceded the formation of larger galaxies such as the Milky Way and Andromeda.

Researchers at Durham University in England have run computer simulations indicating that many of the oldest stars in the Milky Way once belonged to prior, smaller galaxies. These ancient stars were torn from their parent galaxies during violent collisions. Consequently, the Milky Way encompasses not just native-born stars but also many adopted orphans.

Tales of the Dark Ages

One of the least understood periods of cosmic history is the estimated 200-million-year span between the recombination era and the formation of the first stars, an era dubbed the "Dark Ages." During that lengthy interval, gravity quietly collected much of the gaseous material in the universe—mainly hydrogen and helium, with a trace of lithium—into increasingly massive clouds.

Theorists speculate that cold dark matter played a major role in the gathering of atoms together. Clumps and strands of cold dark matter could well have provided the skeleton upon which ordinary atoms could cling. While the leftover cosmic background radiation uneventfully continued to cool—shifting to the invisible portion of the spectrum—space patiently waited for the more dramatic glow of starlight. Until the critical mass for fusion was reached, none of the fledgling orbs could combine its hydrogen in a thermonuclear reaction.

Astronomical instruments such as the Hubble Telescope offer little glimpse of the era before galaxies. The older the objects sought, the farther out telescopes need to scan. Peering farther and farther backward in time presents an increasingly formidable challenge. Consequently, much of the speculation about the Dark Ages of the universe derives from sophisticated computer simulations. These rely on models of cold dark matter clumping together along the lines of the primordial density ripples and coalescing into halos. Hydrogen and other gases then condensed around the denser regions of such dark matter structures.

To form stars, the primeval gas clouds needed to cool. Otherwise their internal pressure would prevent them from contracting enough to fuse and shine. Later in the history of the cosmos, after the first generation of stars was born and died, higher elements would absorb photons and serve as coolants. Such energy absorption allowed gases to settle into tighter arrangements and become the next generation of stellar bodies. In the Dark Ages, however, no such higher elements existed. Atomic hydrogen couldn't do the trick, as whenever one atom released a photon, another would pick it up, volleying energy particles back and forth like a cosmic pinball game. Such bouncing of photons would keep the gas clouds hot and amorphous rather than cooling them into a globe. The key, instead, was molecular hydrogen, with two tightly bonded hydrogen atoms per molecule. Though uncommon, there was enough molecular hydrogen to absorb enough photons over time to serve as the primary coolant. Gradually the gases settled into dense enough bodies to start the fusion process and become the first generation of stars, called Population III.

Generations of Stars

Astronomers count backward in time and assess the portion of higher elements—which they call "metals"—to label stellar generations. "Metals" in astronomy has a different meaning than in chemistry, as it includes any element other than hydrogen and helium. Carbon, nitrogen, and oxygen, for example, are considered metals in astronomy (but not in chemistry). In order of appearance, containing successively higher percentages of metals, the three stellar generations are called Populations III, II, and I. Note the reverse order, reflecting how in astronomy farther away means increasingly backward into the past.

The youngest generation of stars, housing a full palette of chemical elements, is called Population I. These "metal-rich" stars are generally found in or close to the plane of spiral galaxies such as the Milky Way and Andromeda. The Sun is an example of a mature Population I star. Population I stars can be produced only when there are sufficient higher elements around to allow them to condense, relatively late in the history of the universe. Most astronomers believe they are the stars most likely to have planets. Certainly, rocky planets such as Earth, rich with iron and other higher elements, can be produced only in systems with such metal-rich stars.

Population I stars were seeded by the catastrophic demise of an earlier, metal-poor generation, Population II. Typically older, dimmer, and cooler, Population II stars are often found in globular clusters (spherical formations of stars in galactic halos) and close to the centers of galaxies. Over their lifetimes they build up in their cores higher elements that are released into space during supernova explosions.

Population III stars, the metal-free progenitors of Population II, thereby represent the stellar grandparents of stars such as the Sun—even simpler and more volatile. Computer simulations indicate that Population III stars were extremely massive—ranging from thirty to three hundred times the mass of the Sun. Each was extraordinarily hot and bright, fervently engaging in the vital process of transforming hydrogen into heavier elements through cycles of nucleosynthesis.

Early simulations of how Population III stars formed indicated that they were solitary bodies rather than binaries (gravitationally connected pairs). However, in 2009, computer studies conducted by Matthew Turk and Tom Abel of the Kavli Institute for Particle Astrophysics and Cosmology in California, along with Brian O'Shea of Michigan State University, yielded the unexpected result that pairs may have been relatively common. The implication of their findings is that twinned Population III stars could have been less massive than solo stars. Instead of hundreds of times the mass of the Sun, each member of a binary system could have been less than one hundred solar masses. The lower mass better matches models of how a range of higher elements formed in their cores and were released in supernova bursts.

With the formation of the first massive stars, the cosmic Dark Ages drew to a close. Its final curtain call was a remarkable change in the nature of the leftover interstellar gases. As the fledgling stars began to shine, their radiant ultraviolet light stripped off the electrons from the neutral atoms in the interstellar medium, turning them back into ions. In other words, the union of protons and electrons into neutral atoms during the epoch of recombination ended (among the interstellar medium) in a kind of divorce. That divisive final stage of the Dark Ages is called reionization.

Today almost all the hydrogen in space is ionized. Therefore one way of establishing when the Dark Ages ended is to look for evidence of neutral hydrogen in the universe's past. One method of doing so involves what is called the Gunn-Peterson effect, proposed in 1965 by Jim Gunn and Bruce Peterson, then at Caltech. Gunn and Peterson showed that neutral hydrogen would absorb certain frequencies of light. Light coming to Earth from a part of space with considerable neutral hydrogen would have those characteristic spectral lines blocked. Using the Doppler effect (a way of measuring the speed of objects), astronomers could then determine the era of the universe when the neutral hydrogen existed. In 2001, a team of scientists working with the Sloan Digital Sky Survey first used this effect to discover evidence of neutral hydrogen from the Dark Ages—dating the atoms to be approximately 14 billion years old.

Voracious Supermassive Black Holes

Research suggests that Population III stars ended their lives explosively in supernova blasts that left behind black hole remnants. The black holes formed from the implosion of the stellar cores into ultracompact objects—gravitationally powerful enough that not even light could escape. Thanks to the bursts, space would have scattered debris, which for the first time would contain elements higher than hydrogen, helium, and lithium. Meanwhile, the black holes would devour any materials within their grasp—a process called accretion—and grow over time. Those that encountered each other would merge into greater bodies. The largest among these could have helped provide the gravitational impetus for galaxy formation.

Astronomers believe that supermassive black holes lie at the centers of most (and maybe even all) galaxies. They range in mass from hundreds of thousands to billions of times that of the Sun. While many researchers believe that such titans arose from the growth and/ or mergers of star-size black holes, they have yet to confirm the existence of intermediate-size black holes. Such middleweights would form the "missing link" between the star-size and supermassive varieties—pointing to an evolutionary process that transforms lighter into heavier black holes.

In 2009, an international team of astronomers headed by Sean Farrell of the University of Leicester in England discovered an extremely powerful X-ray source that could possibly represent an intermediate-size black hole. Called HLX-1, it is the most luminous X-ray beacon ever found. Further studies, in 2010, with the Very Large Telescope at the European Southern Observatory in Chile, confirmed its location as the spiral galaxy ESO 243–49, approximately 290 million light-years from Earth. Although analysis of the data is still ongoing, and a definitive conclusion has yet to be reached, researchers are optimistic that they may have found a long-sought example of a gravitationally collapsed object of middling size. As Farrell remarked,

> This is very difficult to explain without the presence of an intermediate mass black hole of between ~500 and 10,000 times the mass of the Sun.[1]

Firm proof that intermediate-size black holes exist would fill an important gap in the story of how the largest ones form. Until then, we are still not certain if star-size black holes can evolve into supermassive black holes, or if another astronomical process is going on—such as the catastrophic implosion of tremendous quantities of material. The centrality and ubiquity of supermassive black holes in galaxies strongly suggests that they played important roles in how galaxies evolved.

Frenzied Flashes: The Mystery of Gamma-Ray Bursts

As if quasars weren't baffling enough, astronomers have had to contend in recent decades with another bewildering class of energetic phenomena called gamma-ray bursts. These are quick, sporadic flashes of gamma radiation that seem to appear from out of the blue—lasting from a few thousandths of a second to several minutes. Gamma rays are the highest-frequency and most energetic form of light.

Anomalous gamma-ray flashes were first recorded the late 1960s, when the U.S. military launched a special series of satellites with gamma-ray sensors designed to detect possible Soviet nuclear testing. Atmospheric testing of nuclear weapons had just been banned, and the Vela spacecraft, as they were called, served to verify compliance with the treaty. From time to time the probe relayed back to Earth evidence of brief, random gamma-ray signals from various directions in space. In 1973, Los Alamos researchers Ray Klebesadel, Ian Strong, and Ray Olson took note of these in a paper published in the prestigious *Astrophysical Journal Letters*, and the search for the origins of gamma-ray bursts began.

Like quasars, astronomers weren't sure at first if gamma-ray bursts represented objects within the Milky Way, or more distant entities far beyond its periphery. Frustrating attempts to pin down their properties was a dearth of signals beyond the gamma wavelengths—for example, in the visible range—that could offer key supplemental information. Such added signals were required to match the profiles of these bursts to theoretical models of astronomical events that could potentially release such energy.

NASA's 1991 launch of the Compton Gamma-Ray Observatory was a vital step forward toward the understanding of gamma-ray bursts. It was equipped with a sensitive instrument called the Burst And Transient Source Experiment (BATSE), designed to record the fleeting imprints of energetic gamma rays from deep space. During its nine-year experimental run it captured the signals of more than twenty-seven hundred gamma-ray bursts. The one thing it couldn't do, however, was collect data, after detecting bursts, from other regions of the light spectrum. Fortunately, the launch in 1996 of the Italian-Dutch X-ray satellite BeppoSAX, which also carried a sensitive gamma-ray detector, would offer such a critical follow-up capability. (It was named Beppo in honor of accomplished Italian physicist Giuseppe "Beppo" Occhialini, a pioneer in the study of cosmic rays.)

By then, the most promising model that researchers wished to test was called the "collapsar" theory of gamma-ray bursts, also known as the "hypernova" idea. Proposed in 1993 by astrophysicist Stan Woosley of UC Santa Cruz, it envisioned ultramassive stars undergoing supernova blasts that failed to eject much of the stellar material. In such a case, while a giant star's core collapsed into a mighty black hole, the extra matter would swirl around it as a mammoth accretion disk. As gobs of stellar substance poured into the black hole at an astounding rate, energy would build up within the dying star's interior. A second blast from the core would send a huge shock wave through the accretion disk at close to the speed of light, blowing off the star's top and bottom like the shattering eruption of Krakatoa. As the shock wave tore through the outer material in a blazing fury, it would send streams of energetic gamma rays pouring into space—recorded billions of years later by distant astronomers as a gamma-ray burst. That wouldn't quite be the end of the story. Like a raging demon, the blast would proceed through interstellar space, interact with dust and other material it happened to encounter, and generate an afterglow of less energetic radiation such as X-rays and visible light. Thus a key component of the verification of Woosley's theory would be the detection of such an afterglow.

In February 1997, an event detected by the BeppoSAX satellite proved the smoking gun for identifying the culprit causing such cataclysmic explosions. Immediately following the discovery of gamma-ray burst GRB 970228 (the number refers to the date), its X-ray detector kicked in and found the first indications of an afterglow. That extra fading signal proved key to determining the burst's location and establishing its enormous distance and brightness. As more such one-two-punches were found, with X-ray blasts following the gamma-ray signals, the collapsar/hypernova hypothesis seemed to be holding up well. Yet scientists wanted even more solid proof. What would really clinch matters would be seeing signs of a distant exploding star in tandem with the gamma-ray and X-ray flashes.

In 2000, the Compton Gamma-Ray Observatory was forced down from the skies—not by an alien race eager to protect the secrets of hypernovas but rather by NASA itself. One of its navigational gyroscopes had failed, and NASA feared that another instrument failure could send it on an uncontrolled crash course to the ground. The agency estimated that in the worst-case scenario there was a one in one thousand chance that its debris would land on somebody and kill that person.[2] Consequently, NASA decided to avoid that risk and bring it down in a controlled fashion. The monitored crash landing over the Pacific caused no known damage.

Later in the same year, to further the study of gamma-ray bursts, NASA launched the High Energy Transient Explorer-2 (HETE-2) from Kwajalein Atoll, part of the Marshall Islands in the Pacific. The location was chosen for its closeness to the equator, as it was thought that (due to the alignment of Earth's magnetic field) an equatorial orbit would minimize the interference of high-energy electrons from space. To enable speedy analysis of bursts as well as their fading afterglow, NASA packed the satellite with a sensitive gamma-ray detector and two X-ray detectors. The system was designed to have a hair-trigger response time. As soon as it sensed a burst's telltale signal it was programmed to determine its location and relay that information to other space- and ground-based instruments for additional verification and study.

NASA's investment proved well worth it when on March 29, 2003, HETE-2 instruments picked up a medley of signals from GRB 030329, the closest and brightest gamma-ray burst detected up to that point. Shortly thereafter, a 40-inch telescope at the Siding Spring Observatory in Australia recorded the optical afterglow, enabling astronomers to show that the burst emanated from an enormous supernova explosion 2.65 billion light-years from Earth. Because of its tremendous importance in unraveling the gamma-ray-burst mystery, establishing without a doubt that they are associated with remote exploding stars, astronomers nicknamed the event the "Rosetta Stone."

Woosley could not hide his excitement. "The March 29 burst changes everything," he said. "With this missing link established, we know for certain that at least some gamma-ray bursts are produced when black holes, or perhaps very unusual neutron stars, are born inside massive stars. We can apply this knowledge to other burst observations."[3]

He and other astronomers working in the field have acknowledged that there are still unsolved mysteries associated with gamma-ray bursts. Particularly baffling is their range of durations: some in milliseconds and others in minutes. Did these represent distinct types of phenomena, or could some other factor be coming into play? For instance, could the angle of the gamma-ray beams with respect to Earth be affecting how long these signals lasted? Along with quasars, supernovas, pulsars, and black holes, gamma-ray bursts have joined the pantheon of astronomical marvels that, while they can logically be explained, still boggle the mind with their awesome powers.

Dragon Quest

Gamma rays, always the hottest part of the light spectrum, have now become one of the hottest astronomical pursuits. With the development of more precise detectors, interest in mapping the gamma-ray sky and understanding the types of sources that produce such signals has skyrocketed. Gamma-ray bursts are far from the only strange beasts associated with that form of radiation. Lurking in their hidden lairs are other bizarre types of fire-breathers.

On June 11, 2008, a Delta rocket carried the Fermi Gamma-Ray Space Telescope (originally called the Gamma-Ray Large Area Space Telescope, or GLAST, and renamed after physicist Enrico Fermi) high into orbit. Its mission was to serve as a kind of Amerigo Vespucci of the high-frequency realm, completing an atlas of the gamma-ray sky. The "Eurasia" of the gamma-ray map is the Milky Way—an unmistakably bright glow sprawled across the center. Beyond that central continent is terra incognita—a haze of unknown sources.

Before the Fermi Telescope was launched, most astrophysicists thought that the dominant contributors to the gamma-ray fog were the energetic geysers of particles spouting from the vicinity of supermassive black holes as they devoured their hapless prey (encroaching stars, for example). As these jets slammed into nearby gases at near-light speeds, they would release floods of gamma rays. Active galactic nuclei—with their gargantuan central black holes gobbling up untold quantities of material—would be especially productive. Astronomers could not discern these sources individually because of their colossal distances. With enough such jets, however, the gamma rays spewed from myriad fonts would blend into a murky background mist. So went the theory.

Observation yielded quite a shock. Analyzing data collected by the Fermi Telescope, a team led by Marco Ajello of Stanford made a startling discovery. He announced at a March 2010 conference that fully 70 percent of the gamma-ray fog beyond the Milky Way could not be explained by active galactic nuclei. What was lurking in the mire to breathe out more than two thirds of the gamma-ray background? In a nod to medieval lore, the researchers dubbed these unknown sources "dragons."

Alas, there is no wizards' cartography shop to supply a "marauders' map" of where these hidden firedrakes might dwell. Scientists need to locate them the old-fashioned way—by inventing and testing theories as to their nature and whereabouts. The Fermi Telescope will help celestial seekers narrow down alternatives so that the dragons' lairs may someday be found. No word yet on whether the researchers will need to rescue intergalactic princesses and princes in the process.

One conceivable source of massive quantities of gamma rays would be the formation of clusters of galaxies. The consequent acceleration of particles during the merger could generate high-energy radiation. Another possibility would be the shock waves caused by supernova explosions bashing into gases and forcing them to emit gamma rays.

Finally, yet another plausible explanation would be the collision of dark matter particles. These might interact with one another, mutually annihilate, and produce gamma rays as by-products. One of the Fermi Telescope's missions is searching for such signs. If dark matter is prevalent in space, it could conceivably be leaving traces behind that we have failed to recognize so far. Naturally, confirming such a hypothesis depends on identifying what dark matter truly is—or at least discerning its basic properties. As they quest for the true identities of cosmic dragons, astronomers have yet another motivation for resolving the riddle of dark matter.

Into the Vortex

We need not look far to find massive producers of gamma rays. Within our own galaxies there are gigantic gamma-ray sources. In 2010, astronomer Doug Finkbeiner of the Harvard Smithsonian Center for Astrophysics revealed the existence of huge gamma-ray bubbles above and below the plane of the Milky Way. The twin bubbles each extend 25,000 light-years from the Milky Way's center, forming the shape of a barbell. Scientists had not seen the bubbles before because their radiation was camouflaged within the gamma-ray fog.

Researchers speculate that the gamma-ray bubbles could be relics of an enormous burst of energy from a supermassive black hole in the Milky Way's center. In the past, the center of our galaxy was likely much more violent and prone to such eruptions. Fortunately, today the Milky Way is mature and not subject to such outbursts.

Someday, perhaps, astronauts will explore the depths of our galaxy and visit its center. Maybe some will be brave (or foolhardy) enough to encroach upon the lair of a supermassive black hole. Unlike an ordinary, stellar-size black hole, a supermassive black hole would offer ample time for an explorer to visit before ultimately being crushed.

Physicist Vyacheslav Dokuchaev, of the Institute for Nuclear Research of the Russian Academy of Sciences in Moscow, has speculated that advanced life could exist inside a supermassive black hole, assuming that it rotates or is electrically charged. Rotating or charged black holes, he has calculated, offer the possibility of stable planetary orbits. Savvy beings could exist on such planets, adapted to the mighty tidal forces that would tear less-suited creatures apart. Astronauts entering a supermassive black hole would surely be astonished to encounter such life forms.

Aside from seeking bizarre new life forms, there would be another motivation for investigating the twisted realms of supermassive black holes. Theory suggests that they could harbor gateways to other parts of space, known as Einstein-Rosen bridges or wormholes. The prospect of finding passage to a distant part of our universe—or even a parallel universe—is one of general relativity's most exciting implications.

12

Can We Journey to Parallel Universes?

Wormholes as Gateways

To know the universe itself as a road—as many roads—
as roads for traveling souls.

—WALT WHITMAN, *SONG OF THE OPEN ROAD*

If scientists eventually establish the existence of the multiverse, intrepid individuals might naturally wonder if they could ever visit parallel universes. Could there be gateways to other realms? Strangely enough, Einstein's general theory of relativity seems to permit such passages, at least hypothetically. In 1935 he and his research assistant Nathan Rosen proposed the notion of connections between otherwise separate regions of space—an idea that has come to be known as Einstein-Rosen bridges or wormholes. You may wonder how a bridge could be the same as a wormhole, or tunnel, but both terms describe a way of using a mathematical connection to span the gap between two otherwise separate regions of space.

Einstein and Rosen's 1935 paper was an attempt to clear up a profound mystery in gravitational theory while attempting to solve another in particle physics. The presence of singularities—points of infinite density—in general relativity greatly disturbed Einstein. He didn't like the fact that the Schwarzschild solution, describing how space-time warps in the presence of a spherical mass, has a gaping central hole—a rip in the very fabric of reality. Space and time simply have no meaning at that infinitely compact place.

The term "black hole" had yet to be coined by John Wheeler, but Einstein already eschewed such a concept. Einstein believed that the universe is governed by a full set of deterministic laws and that these must exclude singular points. When instructing his research assistants about viable solutions, he would often express his feelings in religious terms. He would explain to them that developing a theory with singularities represents a transgression against what must be a perfect set of equations set out by the divine to describe the complete behavior of the cosmos.

To remove the singularity, Einstein and Rosen found a way of mathematically continuing the Schwarzschild solution to another "sheet" of the universe—which could be construed as a parallel universe—thus making the central gap a passageway rather than an end point. Instead of a funnel to nowhere, the new solution looked like an hourglass. Anything entering the top half, representing our universe, would pour through the connecting throat and empty into the bottom half. From the lower portion the substance would exude into the second sheet's parallel universe.

Einstein and Rosen saw such bridges between different sheets as a natural way of explaining how elementary particles arise. They postulated that each bridge represents a single particle such as an electron and that multiple bridges delineate interacting particles. Electric field lines thread through the bridges, denoting the way that charged particles produce such patterns. The theory couldn't explain many aspects of subatomic physics, such as why there are so many different kinds of particles—possessing various masses, spin states, and other types of properties—but Einstein and Rosen saw the theory as a start rather than a final product. Because of its lack of predictive power, however, few physicists took their theory of particles seriously.

Starting in the late 1950s, Wheeler revisited the Einstein-Rosen bridge theory, choosing to call the connections "wormholes." He liked to think in pictures, and imagined these as the tunnels bored through apples by sinuous pests. Worms could use these as shortcuts from one side of the apple to the other without needing to wriggle along its round surface. Like Einstein, Wheeler was intrigued by the idea of producing matter from pure geometry, and viewed connected spaces as one of the important elements. Wheeler went beyond Einstein, though, in pondering the use of wormholes for instantaneous transport from one part of space to another.

In a 1962 paper with Robert Fuller of Columbia University, called "Causality and Multiply Connected Space-Time," Wheeler wondered if such instant connections could violate the law of cause and effect. The authors picture a wormhole that connects two distant parts of the universe that would take much longer to journey between by means of ordinary space travel. If a beam of light passes through the wormhole's throat it could outpace the conventional speed of light. In that case, the wormhole could convey an effect before standard communication delivers its cause. Such a reversal of the normal order of events would represent an unacceptable breach of the law of causality.

For example, consider two planets, Gustav and Holst, that communicate with each other by two different message services: a wormhole gateway for emergency communications and conventional radio signals for other types of messages. The two planets are 4 light-years away from each other, meaning that the conventional signals take 4 years to span the distance between them. Therefore messages sent via wormhole easily outpace the speed of light, allowing for bizarre reversals of the order of cause and effect.

Imagine that in the year 3000 a madman on Gustav named Doctor Destructo sends a message to Holst via conventional radio. The evil doctor announces that he is going to set off a devastating explosion on Gustav that destroys the Holst embassy unless the authorities on Holst transfer a certain amount of money into his account. In the broadcast, he shows how he has started a timer for a one-year countdown. A bomb will automatically detonate in 3001 unless he receives the funds. Only once he is sure the money is in his account will he stop the clock. In his delusional state, the doctor forgets about the 4-year time delay for conventional messages. He waits a year, then—frustrated by the lack of response—lets the detonation occur. News and images of the catastrophe are sent to Holst via the emergency wormhole channel. Its denizens wonder what the reason was for such an abhorrent act. They have seen the effect but don't know the cause. Then, in 3004, the threatening message from the doctor finally arrives, including the image of him starting the timer for the one-year countdown. The people of Holst experience the effect (the explosion) before experiencing the cause (the unanswered message). In marked contrast to familiar experiences, causality has taken place in reverse order.

In their paper, Fuller and Wheeler offer good reason why such a violation of causality could not occur. They demonstrated that any signal or material trying to enter a wormhole based on an extension of the Schwarzschild solution would cause its throat to become unstable and close off. Consequently, unless future engineers could find a means to prop the throat open, a Schwarzschild wormhole, such as a black hole connection, would immediately become blocked. No faster-than-light signals could then be transmitted, and causality would be rescued.

Making Contact

Even if the throat of a wormhole within the interior of a stellar-size black hole could be propped open and prevented from shutting down travel, no sane astronaut would dare attempt the journey. Black holes are far too dangerous to attempt passage. Suppose an intrepid voyager tried to enter one in hope of finding a portal to another part of space. Within seconds of passing through the event horizon (the black hole's point of no return), enormous tidal forces would stretch, crush, and pulverize the unfortunate soul. It would take a foolhardy explorer indeed to try to experience first-hand a black hole's innermost secrets.

Supermassive black holes, much larger and far heavier than their stellar kin, would offer considerable more time for reflection on one's fate. Intrepid voyagers entering such a monstrosity would have ample time to explore before being smashed to bits. If they are lucky, and the black hole is rotating (a general relativistic solution derived by Roy Kerr), there would even be a way of avoiding the singularity. Rather than a point, a rotating black hole would have a ring singularity. By steering away from the crushing loop, the explorers would have the leisure of trying to search for a wormhole that could potentially (if somehow stabilized) allow them to escape. However, they would still face the peril of a rain of deadly radiation due to the influx of material from nearby stars absorbed by the powerful body. Thus, while it would be a longer, potentially survivable (with sufficient protective gear) journey, it certainly wouldn't be pleasant. No sensible astronauts would be likely to include such expeditions in their travel plans.

In 1988, interest in traversable wormholes experienced a dramatic revival when Caltech physicist Kip Thorne and his graduate student Michael Morris published a groundbreaking paper on the topic. Having received his own PhD under Wheeler, Thorne was well versed in the subject of black holes and wormholes. He was well aware of all the difficulties associated with Schwarzschild wormholes and all the dangers that black hole travel would pose to potential space voyagers. Therefore, when Thorne's friend the noted astronomer and writer Carl Sagan mentioned the idea of using black holes as a plot device for his novel *Contact* (to allow for interstellar travel), Thorne steered him away that idea and instructed Morris to investigate a more workable solution.

Morris-Thorne traversable wormholes offer a distinct general relativistic solution specifically designed to avoid the problems that plagued conjectured black hole journeys. If such wormholes could be found or engineered, travelers would enjoy safe, swift passage without fear of decimation. Tidal forces and radiation would be kept to a minimum within the throats through which the explorers would pass, enabling them to voyage comfortably from one mouth of the wormhole to the other. These mouths could be in different parts of our own universe, or in separate universes altogether. Thus a traversable wormhole could conceivably allow access to a parallel universe.

Propping open the throats of such traversable wormholes would be a hypothetical substance called "exotic matter." Exotic matter would have the unique property of negative mass, offering a kind of negative pressure or antigravity. It is strange to think of mass as having a negative value; all known particles and even antiparticles have positive mass. If you walked into a deli and asked for a negative pound of salami, you'd probably get strange looks. Yet, as we've seen in our discussion of dark energy, the vacuum itself exhibits negative pressure. Therefore the existence of exotic matter is not as far-fetched as it would seem.

Morris and Thorne speculated that an advanced civilization could potentially mine the vacuum for the material needed to create wormholes. They imagine such a society scooping up some of the quantum foam—with its froth of connections continuously bubbling due to quantum uncertainty—and enlarging a microscopic wormhole into the size needed for passage. Thus, as in the case of inflation blowing up minuscule fluctuations into large-scale structures, a technologically savvy future culture might enlarge a minute quantum connection into a macroscopic wormhole.

Subsequent work by physicist Matt Visser of Victoria University of Wellington, New Zealand, generalized Morris and Thorne's work into a variety of traversable wormhole configurations. Visser showed how the amounts of exotic matter needed to form a wormhole could be minimized by fashioning them using particular kinds of geometries—such as flat mouths with the exotic matter only on the edges. Explorers could travel through the flat entranceways and thereby avoid the energetic influences of the peripheral zones.

Reportedly, Thorne has written a plot for an upcoming movie about wormhole space travel, called *Interstellar.* Given that *Contact* was made into a movie, it would be the second film directly related to his traversable wormhole proposal. There have been numerous other treatments in various media. For example, *Star Trek: Deep Space Nine* was a 1990s television series based on wormhole travel. Wormholes have become a cultural icon as well as an intriguing physical construct.

The Trouble with Time Travel

With the advent of improved wormhole models with enhanced safety features and noncollapsible throats, the causality issue has reared its backward clock face once more. In 1989, Morris, Thorne, and Ulvi Yurtsever developed a means by which wormholes could be modified for use as time machines, whisking astronauts backward in time. They imagine taking one of the mouths and speeding it close to the velocity of light. The other mouth would remain at rest, relative to an advanced civilization's home planet; let's call it Chronos. Einstein's special theory of relativity mandates that objects moving at near-light speeds would age more slowly relative to the perspective of a stationary observer.

Therefore, if the moving mouth is traveling quickly enough, it might age only 1 year, while for the stationary mouth and Chronos a full 100 years had passed. If the wormhole were constructed in 3000, then the calendar year would be different on each side. While the year would be 3100 near the stationary mouth, it would be only 3001 near the moving mouth. Now suppose a nostalgic centenarian astronaut named Tempra, born in 3000 and living on Chronos, wishes to revisit the days of her youth. She need only journey into the stationary mouth, emerge via the moving mouth to travel backward in time to the year 3001, and voyage home to Chronos, and she could relive her salad days.

We can readily see how such backward time travel could create causality paradoxes. If Tempra manages to return to Chronos in only five years, let's say, it would still only be 3006. She could search for her childhood home and encounter herself at age six. She could even advise herself not to become an astronaut, but rather to become a high-powered investor instead. She could tell her younger version to dump out her piggy bank and invest in United Wormhole Enterprises—which will be a leading corporation in the future. Suppose her younger self does just that and becomes a wealthy philanthropist rather than a space explorer. Then how did she go back in time and offer herself advice? The cause of her life-changing decision would be a phantom.

Even if Tempra didn't meet her younger self, she could still create paradoxical situations. If she wasn't careful, she could reveal secrets of the future that would change history. For example, she could warn a leader about an assassination plot that she knew would happen. However, the ramifications of that leader surviving could influence history in unpredictable ways. For example, he could end up starting a war that wouldn't have happened otherwise. What if that war ended up disrupting the space program, among other ramifications? Once again, Tempra's backward trip through time would be inexplicable. A contradiction would occur between two different narratives—one with her making the journey and the other with her unable to do so because of the chain of events that she created herself by her time-traveling actions. It is clear, therefore, how backward time travel could create a paradoxical oscillation between two clashing realities. Such dilemmas are sometimes dubbed the "grandfather paradox," after the possibility of killing one's grandfather and preventing oneself from existing, and the "butterfly effect," after Ray Bradbury's speculative scenario of someone traveling backward to the age of the dinosaurs, inadvertently stepping on a butterfly, and setting off a domino-like reaction that disrupts the course of history.

We emphasize that such paradoxes are associated with past-directed travel, not all attempts to journey through time. Purely future-directed travel would be physically acceptable and paradox-free. Special relativity would allow astronauts to travel indefinitely into the future, as long as they had the technical ability to speed up sufficiently close to light speed. As they approached the speed of light they would age more and more slowly compared to those they left behind on Earth. They might return to Earth and find that a hundred—or a thousand—years had passed. Even so, they would not violate causality, as they would influence only events that were yet to happen. If the future is unwritten, and the time-traveling astronauts change it, they would not create a paradox.

Past-directed travel, where the chain of causality could be reversed by influences from later times to earlier ones, would be a different story. The distinction is something called closed timelike curves (CTCs): strands of reality that intersect with themselves backward in time. Surprisingly, there is nothing in general relativity that prevents CTCs from forming. Aside from wormholes, other general relativistic solutions have been found that contain CTCs, and the possibility of time paradoxes such as meeting oneself and disrupting the flow of history.

There are a number of possible solutions to such time travel paradoxes. Russian physicist Igor Novikov has conjectured a self-consistency principle that mandates that CTCs are allowed only if they do not lead to contradictory situations. For example, people would not be able to travel into the past and prevent themselves from existing. On the other hand, they could journey backward in time and hand themselves the blueprints for a time machine that would allow them to voyage back in time. The closed loop they created in that case would be self-consistent. The example he used was a billiard ball going back in time, emerging on a table, and knocking itself into a wormhole—which brings it back in time once again. The loop would continue, without contradiction, ad infinitum.

Note that for Novikov's scheme to be valid, human action would need to be at least somewhat deterministic, or, alternatively, reality would need to account for all the choices someone might make. In a fully mechanistic scheme, a time traveler would lack the free will to change history but, without realizing it, would simply be following a script. As an alternative, a time traveler would try to alter the past, but no matter what he did, the chronology of events would be indelible. For example, someone could strive to prevent Lincoln's assassination, but somehow be prevented from entering Ford's Theatre. If he finally manages to force his way in, the ensuing disturbance might allow Booth to perform his heinous deed anyway. Admittedly, the idea that every attempt to change history would lead to a counteraction preventing such a move is rather far-fetched; however, that hasn't stopped it from being a plotline for numerous speculative stories.

Other physicists have suggested that the fundamental principles of physics would preclude CTCs from forming at all. For example, in a 1992 research paper, "Chronology Protection Conjecture," Stephen Hawking proposed that "the laws of physics do not allow the appearance of closed timelike curves."[1] He based his hypothesis on two aspects of CTCs. First, the boundary between CTCs and ordinary parts of the universe would violate a principle called the weak energy condition, which mandates positive values of energy and mass. CTCs, if they existed, would constitute regions so distorted that the direction of time and space would be swapped. Instead of moving through space, people could hypothetically travel through time. Each such topsy-turvy domain would be separated from normal space-time by a region called the Cauchy horizon—a kind of invisible barrier in which space and time switch roles. The problem is, as Hawking pointed out, this frontier would appear to have negative energy density, breaching the weak energy condition. Given that exotic matter would have negative mass, and mass can be converted to energy, the result was not surprising.

Hawking conceded that quantum gravity could possibly permit negative energy density under certain circumstances. The Casimir effect—an attractive force due to quantum vacuum effects—offers a good example. However, Hawking calculated that the potential completion of a loop in time would produce an energetic "back reaction" that would prevent it from cementing together. Hence the laws of physics would conspire to prevent time tourists from trying to visit ancient Rome and crowd the Colosseum, among other potential CTC vacations. And, as Hawking's fellow erstwhile *Star Trek: The Next Generation* guest star James "Scotty" Doohan used to say, "You cannot change the laws of physics!"

Yet another proposed resolution to the paradoxes created by backward time travel is the possibility of alternative realities, as expressed in parallel universes. For example, Alexander Vilenkin's theory of eternal inflation, spurred by Andrei Linde's notion of "bubble universes," envisions the possibility of myriad universes parallel to our own. As we'll discuss, a version of quantum theory called the Many-Worlds Interpretation also posits alternative realities. If someone traveled backward in time it is conceivable that he would end up in another branch of reality and eliminate the possibility of contradicting the former history. For instance, if a time traveler managed to prevent Lincoln's assassination, he would end up in a timeline in which Lincoln served out his second term. Even if the traveler found himself in a reality in which he had never been born (by inadvertently ruining his parents' courtship, for example, and thus preventing his own birth), there would be no paradox if being born in another branch were allowed. It would be strange to be born in an alternative reality, but not paradoxical, if such parallel strands turn out to exist. It would be in some ways like being born in Yugoslavia or Czechoslovakia, countries that no longer exist, except that you would have much more explaining to do and couldn't possibly recover your birth certificate unless you had brought it with you.

Life in Other Universes

It is intriguing to imagine life in parallel universes. If the physical laws and conditions in other universes were similar to ours, they could potentially house myriad habitable worlds. But what if conditions were very different?

The search for extraterrestrial life has gained considerable momentum in recent years with the discovery of hundreds of planets beyond the solar system. Since 1995, various teams of astronomers have measuring the wobbling of stars using the Doppler effect and determined the properties of any orbiting planets. Most of these newly discovered worlds are much larger than Earth. Many orbit in scalding proximity to their parent star, offering little chance of life as we know it.

However, recently there have been improved prospects for habitable planets. The discovery of a rocky planet circling the star Gliese 581, 20 light-years away, set headlines blazing with the announcement that it could have the right conditions for life. Specifically, although the star is a faint red dwarf, and the planet orbits it very closely, the combination of dimness and proximity offers the possibility that some regions of the newly found world have temperatures moderate enough to support life. Those prospects were exciting enough for codiscoverer R. Paul Butler to label it the "first Goldilocks planet."[2]

The Goldilocks zone is the region near a star that is temperate enough (at least according to Earth standards) for there to be reasonable odds that a planet within it might be habitable. Being not too hot and not too cold doesn't guarantee life but just improves the chances. Future studies of rocky planets that analyze spectral lines in their atmospheres would reveal even more information about the possibility of organisms being able to survive there. Finding oxygen and water vapor in a rocky planet's atmosphere would be a giant step forward.

With the search for life in space only in its infancy, perhaps it is premature to envision life in other universes. Nevertheless, contemplating whether other spaces could harbor life is a way of reflecting on how special our universe is. Is ours a Goldilocks universe—uniquely suited for life—or is it just one of many, like the offspring in "The Old Woman Who Lived in a Shoe"?

What if, among the paper-doll-like array of universes, there were domains with completely different physical laws? If the laws differed enough from ours, some scientists would argue that they wouldn't be able to support worlds with life. For example, if the strong nuclear force were much less potent, stable higher nuclei, such as the carbon and oxygen necessary for life, wouldn't be able to form. Such is the reasoning behind the Anthropic Principle's pinholed sieve, which permits only very specific conditions for universes with life and hence justifies, because we are here, why these conditions must be just so.

However, MIT physics professor Robert Jaffe, postdoctoral researcher Alejandro Jenkins, and graduate student Itamar Kimchi have recently demonstrated that certain physical laws in another universe could differ from ours yet still support the possibility of life on planets there. They showed that even in a universe where the masses of elementary particles are greater or less than in ours, it is conceivable that the elements required for life could form nonetheless. The key, they suggested, is to look for combinations of particles that, while unlike what we are used to, could still produce stable variations of hydrogen, oxygen, and carbon. These alternative versions could combine to form organic molecules, water, and other ingredients for life. As Jenkins remarked about altering particle masses,

> You could change them by significant amounts without eliminating the possibility of organic chemistry in the universe.[3]

For example, suppose in another universe protons were slightly more massive than neutrons, unlike in ours, where protons are somewhat lighter. This could transpire if the down quark, which in our universe is heavier than the up quark, were lighter instead. Up and down are the two most basic quark flavors, or categories. Neutrons have one up and two down quarks, and protons have two up quarks and one down. Therefore a change in quark masses could bolster protons in relationship to neutrons. If that happened, while ordinary hydrogen (one proton) would become unstable, it is possible that an isotope such as deuterium (one proton and one neutron) or tritium (one proton and two neutrons) could become stable and serve as a replacement. Similarly, isotopes of oxygen and carbon (such as carbon 14) could become stable, serving as the basis for new types of long, stringy organic molecules.

It is as if the Three Little Pigs found themselves in a town where the bricks were oddly shaped and the mortar was too chunky to spread. The wolf was lurking and they needed an abode. Realizing that they couldn't build stable walls using those elements, they could give up and run crying to Goldilocks that they couldn't make a house that is just right. Or they could look around and find stones and grit that were of suitable proportions and consistency to construct sturdy walls. They could construct a perfectly suitable house that way. It would have a different texture and appearance but serve just the same. Likewise, in a universe with heavier, unstable protons, it is conceivable that different building blocks could be used for life. More massive protons would not necessarily mean a dearth of planets where life could exist.

Jaffe and his collaborators refer to types of universes that could possibly harbor worlds that could harbor life as "congenial." Tapping into their extensive knowledge of particle properties, they have conducted a thorough investigation of which combinations of altered quark masses could produce congenial universes and which could not. He wondered if nature could be tinkering around itself in seeing which mixtures might be viable. Jaffe said,

> Nature gets a lot of tries. The universe is an experiment that's repeated over and over again, each time with slightly different physical laws, or even vastly different physical laws.[4]

The work of Jaffe and his team has broadened astronomical understanding of what the minimal ingredients for life are, just at a time when researchers seem on the verge of discovering worlds where life could exist in our own cosmic neighborhood. Perhaps visionary sixteenth-century Italian thinker Giordano Bruno, who was burned at the stake for blasphemous remarks (including belief in a plurality of worlds, among other alleged heresies), was a tad conservative in his estimate that *our* universe contains innumerable planets with life. He could have thought even bigger and considered all the other universes, too!

13

Is the Universe Constantly Splitting into Multiple Realities?

The Many-Worlds Hypothesis

Do I contradict myself?
Very well then I contradict myself;
(I am large, I contain multitudes.)

—WALT WHITMAN, *SONG OF MYSELF*

General relativity is not the only branch of physics that offers the tantalizing prospect of parallel realities. Quantum mechanics describes situations in which particles are in states that are a combination of two possibilities. For example, electrons have two types of spin states, called "up" and "down." We can think of these as tops rotating either counterclockwise or clockwise. Electrons don't actually rotate, but when in a magnetic field their behavior mimics spinning charges. Their spin can point either in the direction of the field—up—or in the opposite direction—down. Strangely enough, before researchers switch on a magnetic field to measure an electron's spin state, quantum theory informs us that an electron is in a superposition of both possibilities. As odd as it sounds, its spin state is simultaneously up and down. Only after a researcher measures that state is it said to "collapse" into one of the two options.

Thinking of an electron in a mixed state seems somewhat abstract. But what about a cat? We can thank Austrian physicist Erwin Schrödinger for the enigmatic, gruesome image of a zombie cat trapped in a kind of limbo between existence and death until a researcher's measurement releases its soul (back to its body or onward to the beyond).

Zombie Cats

In the history of quantum physics, Schrödinger was a paradoxical figure. Although he was honored with the Nobel Prize for his pivotal contributions to quantum mechanics, he spent much of his life challenging the ramifications of that enterprise. He never liked the idea of wave-particle duality in which wave functions spontaneously collapse into specific configurations corresponding to certain particle properties. He and Einstein shared a disdain for random "jumps" from one quantum state to another—preferring instead smooth, predictable transitions within the structure of a deterministic wave equation.

There were other combinations that Schrödinger found distasteful. According to his biographer Walter Moore, a pairing of food and drink offered to him at a New York dinner during Prohibition helped trigger a lifelong aversion to the United States. As Moore remarked,

> An important factor in Erwin's intense dislike of the American way of life was the "great experiment." An occasional glass of good beer or bottle of fine wine would doubtless have made everything seem more bearable. With a plate of succulent Chesapeake Bay oysters he was offered a choice between sweet ginger ale and chlorinated ice water. "To the devil with Prohibition," he exclaimed.[1]

Although he was averse to oysters served with tap water, Schrödinger did not object to unusual medleys of relationships—which his wife, Anny, seemed to tolerate. In the early 1930s, he had a sexual liaison with Hilde March, the recently married wife of his research assistant Arthur

March. After Hilde became pregnant and gave birth to their daughter Ruth, she became something like a second wife. Schrödinger's amorous attentions continued to remain in a mixed quantum state, shared among several women at once. He maintained numerous other affairs throughout his life.

In 1934, when offered the prestigious Jones professorship of mathematical physics at Princeton, both predilections (epicurean and nuptial) may have come into account in his decision to decline. Apparently the oysters-and-ice-water incident had left him with a bad taste for American culture. Accustomed to the bohemian mores of interwar Europe, could he move to a country where an austere ban on alcohol had just been lifted?

Moreover, Schrödinger did not like the reaction he got to his complex marital situation. He reportedly discussed with President Hibben of Princeton the possibility that both Anny and Hilde would join his household, to help take care of his (then expected) daughter. According to the story, Hibben's response was far from positive. Subsequently Schrödinger was worried that if he moved to Princeton he could even be prosecuted for bigamy.[2] These issues, along with salary concerns, may well have played a role in his choice to stay in Europe.

Ambiguity continued to follow Schrödinger like mist to a mountaintop. He often couldn't make up his mind about life choices and seemed to want things both ways. After returning to Austria, he spoke against the Nazi regime when it was a threat across the border. Then, after Austria's annexation by Hitler, Schrödinger wrote a letter asserting that he had changed his views in favor of the Nazis. However, he ultimately decided to flee Austria for Ireland, where he dismissed the letter as a misguided attempt to protect himself and keep his job.

Today, when most people think of Schrödinger they think of his cat. Nothing could be more ambiguous than that phantom feline poised with two paws on each side of the pearly gate! That teetering tabby is etched into the scientific psyche as a half-living, half-dead monstrosity. On the face of it, its eerie fate is more suitable for ghost stories than for the pages of stately journals. Yet that clawed symbol of quantum uncertainty has spurred a rethinking of space and time that has led to the concept of the multiverse. Though that zombie cat may (or may not) be long gone, its influence purr-sists.

The story of Schrödinger's cat began as a response to a paper by Einstein and two of his research assistants. In 1935, Einstein, Boris Podolsky, and Nathan Rosen published an article, "Can Quantum-Mechanical Description of Physical Reality Be Considered Complete?," intended as a rejoinder to the Copenhagen (standard) interpretation of quantum mechanics. Commonly known as the EPR paradox, it imagines a situation where knowledge of a quantum state seems to be instantaneously transmitted over an indefinitely large distance. (This is the "spooky action at a distance" or quantum entanglement mentioned in chapter 9 as a possible explanation for the great voids.)

A simple way of looking at the EPR paradox is to imagine two electrons emitted from the ground state of an atom. From the Pauli exclusion principle, a rule that prevents electrons from crowding too close together, we know that these electrons cannot have all the same quantum numbers and must have opposite values of the quantum parameter called spin. There are two types of spin for electrons, up and down. So we know if one is up the other is down, like kids on opposite ends of a seesaw. However, unless we take a measurement, quantum uncertainty informs us that they are in a mixed state and we don't know which is up and which is down. This is hard to picture, but we can try to imagine a seesaw moving so blurrily fast that we don't know which side is up.

Let's assume that both electrons are released at once, in opposite directions. As they separate more and more, we still don't know which electron has which spin. Now we take a measurement of one of the electron's spin states. According to the Copenhagen interpretation, the electron being examined immediately collapses (with a 50–50 likelihood) into either an up or a down state. Right away we measure the other electron's spin state, and unwaveringly it is the opposite of the first. How does the second electron instantly know what the first one has "decided"? Einstein, Podolsky, and Rosen thought they had slung an arrow into the heart of the theory, but quantum theory survived much stronger than ever after experiments showed that this is what actually would happen.

Schrödinger was most intrigued by the EPR paper's implications and decided to explore them on his own. He wrote a philosophical article, "The Present Situation in Quantum Mechanics," which included an anecdote about an imaginary experiment with a cat that has a chance of either living or dying based on a quantum outcome. In Schrödinger's words, here is what happened to it:

A cat is penned up in a steel chamber, along with the following device: . . . in a Geiger counter there is a tiny bit of radioactive substance. . . . Perhaps in the course of the hour one of the atoms decays, but also, with equal probability, perhaps none; if it happens, the counter tube discharges and through a relay releases a hammer which shatters a small flask of hydrocyanic acid. If one has left this entire system to itself for an hour, one would say that the cat still lives *if* meanwhile no atom has decayed. The psi-function [quantum state representation] of the entire system would express this by having in it the living and dead cat . . . mixed or smeared out in equal parts.[3]

In other words, release of enough poison to kill a cat would be connected with the radioactive decay of a single atom, as measured by a Geiger counter. The cat lives or dies in conjunction with a quantum event that has 50–50 odds. However, the determination of the result of that event happens only after an observer measures it. Until then, the cat is in a juxtaposition of the two possibilities—an equal combination of poisoned and safe.

Schrödinger could not believe that a cat could remain in a mixed state; either it is alive or dead. The standard prediction that the collapse into one of the two possibilities would occur only *after* the lid was opened and the measurement was taken seemed to him most peculiar. Therefore, he thought, by implication, that since any quantum process could be linked to the fate of a cat (or another living creature), it also must be in one state or another. Rather than mixed states, there must be some barrier to knowledge that could eventually be overcome through more powerful theories.

Schrödinger's bizarre result did little to rattle the quantum community—at least at first. The standard interpretation, though fundamentally mysterious in its mechanisms, has enjoyed enormous predictive success. However, not far from the Princeton address where Einstein spent his final years, a younger generation of physicists would embrace the challenge of trying to rethink the premises of how quantum processes work.

House of Mirrors

As a professor and research adviser at Princeton in the 1940s and 1950s, John Wheeler encouraged his students to find innovative, often radical solutions to seemingly intractable problems in physics. Having worked with Bohr, and being a neighbor and friend of Einstein, he was cognizant of the great rift between the probabilistic Copenhagen interpretation and determinism, and of the gap between quantum theory and general relativity. Wheeler sought ways to bridge the divide by imagining hybrids between the two theories—paving the way for the discipline known as quantum gravity.

Wheeler's most prominent student was Richard Feynman. Feynman had a knack for finding practical solutions to problems and presenting them in a way easy to visualize. If ever there was a jack of all trades, he was the one. Not only was he a brilliant researcher, he also would later became a groundbreaking physics educator. He played bongo drums in bohemian clubs and dabbled in painting. His mind constantly raced toward new challenges—which he almost inevitably would master. Nature for him was like an elaborate puzzle for which the rules weren't laid out but needed to be discovered. He would handle each step like a crossword aficionado until he unraveled every clue and filled in each blank. During the Second World War he played a prominent role in the Manhattan Project, where he amused himself figuring out how to crack open safes. Clearly the secrets of nature were no match for such a keen problem-solver.

In 1948 Feynman decided to tackle a daunting problem in physics: the quantum interaction between charged particles, such as electrons, through the exchange of photons. He set to work developing a shorthand notation—now known as "Feynman diagrams"—to describe what was going on. Like a football coach planning a strategy, he characterized each player with arrowed lines (representing electrons) or squiggles (denoting photons), showing all the possibilities for electrons meeting up and tossing photons back and forth. As crazy as it seemed to him, he found that by sketching out every conceivable scenario that fit the conservation laws and assigning weights to each, he could calculate the likelihood of certain outcomes. In other words, reality was a weighted sum of all possibilities. He formulated this in precise mathematical language to express quantum mechanics called "sum over histories."

Feynman's technique was one of the first mathematical expressions of the idea of parallel realities. Unlike classical trajectories, by knowing the initial conditions and final conditions for an interaction between particles, you couldn't assume that they took only one path to get there. Rather, all roads are taken as long as they are physically viable. Each route is assigned a probability value, with the classical path turning out to be the most likely.

We can understand the difference between the classical path and the quantum web of trajectories by imagining teams of explorers trying to take the shortest route through a mountainous terrain. Each of the classical explorers, ironically enough, has a special GPS system that tells him exactly which path to take to minimize the distance. It is like Newton in a *nüvi* (portable GPS)—calculating the trajectory with precision. The quantum explorers, on the other hand, each have a crude terrain map. Most learn to figure it out and adhere close to the shortest path. However, some outliers, less adept at map-reading, take other, more circuitous routes. The result is a fanning-out of trajectories, with the most probable one corresponding to the shortest path. If a team of doctors comes along to assess the conditions of the teams of explorers after their journeys, they would know the classical group's amount of wear and tear with great certainty, but for the quantum group they would have to "sum over histories" by obtaining a weighted average of the amount of stress and strain associated with each of the possible routes.

Feynman did not intend his method to represent actual parallel realities; he was too practical for that. Rather, he meant it as mathematical shorthand for how quantum measurement must take into account the principle of uncertainty. The roads all taken represented the lack of knowledge within our world, rather than bifurcations into alternative worlds. However, another of Wheeler's students, Hugh Everett III, would make that leap.

Everett had an exceedingly brief career as a theoretical physicist—principally consisting of his graduate work at Princeton. Yet, long after he received his doctoral degree and switched into military research, his meteoric contribution continues to dazzle. His 1957 PhD thesis, "On the Foundations of Quantum Mechanics," proposed a solution to

Schrödinger's cat paradox and similar quandaries about mixed quantum systems that enables a complete description without reference to the actions of observers. He conjectured that each time a subatomic event occurs—whether it represents decay, scattering, absorption, or emission—the universe bifurcates into parallel realities. Not only does the quantum interaction split into distinct realities, everything else does, too. Hence anyone observing a quantum experiment would witness a result that depends on exactly which version of truth he or she happens to be in. A mad scientist conducting the cat experiment, for instance, would divide into two parallel selves immediately after closing the lid of the potentially deadly device. One self would open up the chamber and discover delightful purring; the other would find only horrific silence. The bifurcation process would be totally seamless; neither version of the scientist would realize it had happened or be aware that the alternative self existed.

Everett showed that his theory was equivalent to the standard Copenhagen interpretation. Instead of probabilistic jumps and collapses, quantum transitions would be completely deterministic. While there would be a maypole of strands at each juncture, each observer would grasp only a single, steady thread of cause and effect.

One open question about whether the two interpretations matched completely was something called the Born rule: a way of calculating the probability of each outcome. The Copenhagen interpretation did this naturally through a matrix (mathematical table) that gave the odds of each possibility. Everett needed to assume that each observer's strand had different likelihoods. (In 2010, physicists Anthony Aguirre, Max Tegmark, and David Layzer would suggest that the fraction of observers clutching each strand matched the Born rule probabilities, leading to a fascinating potential resolution of the question.)

Wheeler encouraged Everett to write a paper describing the theory, added his own positive assessment, and sent a copy to Bryce DeWitt, another quantum gravity pioneer. DeWitt was involved in planning the first major American conference on general relativity (organized by Cecile DeWitt-Morette in Chapel Hill, North Carolina), and, as editor of its proceedings, accepted Everett's paper. Feynman and Wheeler attended the conference, held in 1957, but apparently Everett did not. Nevertheless, Everett's paper was published in the proceedings.

DeWitt's initial reaction to Everett's idea was supportive of the physics behind it but skeptical about the implications of branching realities. Nothing in our experience tells us that reality can shatter into countless shards. Everett responded by emphasizing how, like Earth's inhabitants not feeling its turning on its axis, nobody would ever sense that something was changing when the bifurcations were taking place.[4]

Eventually DeWitt came to see great merit in the theory. He became the principal torchbearer for the concept, assigning it a catchy name: the Many-Worlds Interpretation (MWI) of quantum mechanics. Despite its profusion of ever-branching realities, he argued that it was the only self-consistent explanation of quantum mechanics that could be applied to the entire universe. Unlike the standard Copenhagen interpretation in which the experimenter causes "collapse" to a particular final outcome, in MWI observers play no role in affecting an experiment. In describing the quantum state of the universe, there are no external observers to measure it and precipitate collapse. Therefore, DeWitt pointed out, by avoiding the need for interactive observers, MWI offered the most objective means of understanding quantum dynamics. As late as a 2002 celebration of Wheeler's ninetieth birthday, he gave a persuasive and well-received talk supporting the theory.

Everett had long passed away by then. He died suddenly in 1982 of a heart attack at age fifty-one. Sadly, his son Mark found him and tried unsuccessfully to revive him, but he had likely been dead for some time. Mark wasn't aware of his father's pivotal contributions until much later in life. He went on to a musical career and is now the lead singer, musician, and songwriter for the Eels.

Until the end, because of his ardent faith in parallel universes, Everett had believed that death was impossible—a philosophy that has come to be known as quantum immortality. Each quantum process, he thought, would lead to a splitting-up of a person's conscious identity. Therefore even if one of the copies dies, others would live on. With further quantum transitions, the surviving copies would then bifurcate again and again in a never-ending progression. Whenever the candle blew out for any of the versions there would always be others left to carry on the flame of consciousness. Hence, Schrödinger's cat would have far more than nine lives!

Occam's Barber Shop

William of Ockham, a fourteenth-century theologian, famously wrote that things should not be unnecessarily multiplied. In philosophy, Occam's Razor (also spelled Ockham's Razor) is the practice of whittling down the set of possible solutions to a problem until the simplest is reached. It advises not to make things more complicated than they need to be. Thus, if you walk outside and see puddles of water, Occam's Razor would suggest that you check if it had rained recently before jumping to the conclusion that a truck transporting piranhas to a nearby aquarium must have sprung a leak after a collision with an escaped rhinoceros.

Cosmology likes to shave down possibilities, too. After all, one of its most famous theorems is that "black holes have no hair," a way by which Wheeler described their simplicity of properties. Therefore, many mainstream cosmologists have become dismayed about the proliferation of multiverse models sprouting long, tangled beards with innumerable strands. Could there be any way of taking them into the salon of simplicity and clipping them down to size?

Each of the parallel universe models has been introduced to eliminate logical gaps in what we observe. For example, the Many-Worlds Interpretation removes the instantaneous, probabilistic jump associated with the collapse of mixed quantum states. Wormholes to other "sheets" help avoid singularities that represent breaches in the fabric of space-time. Eternal inflation is a consequence of the inflationary model—itself a scheme to help explain how the very early universe transformed from chaotic to smooth (overall flat and isotropic). To include yet another parallel universe notion, Leonard Susskind's fitness landscape scheme (discussed in chapter 5) is an attempt to understand how the multitude of string theory vacua could result in a single, comprehensive theory of the natural forces.

With the goal of simplicity as well as comprehensiveness, scientists have sought to find connections between the various types of parallel universe models. Such studies are in preliminary stages, given that the study of quantum gravity, the attempt to link quantum mechanics with general relativity, remains speculative. One step forward occurred at about the time of Wheeler's ninetieth-birthday celebration, when Max

Tegmark presented a hierarchical scheme for grouping parallel universes into various categories, listed as Levels I to IV.

According to Tegmark's taxonomy, Level I represents parts of space beyond the cosmic horizon that are impossible to observe. Thus any regions of the physical universe outside of the observable universe would fall into this class. Among the four levels, Level I would be the most widely accepted, given that almost all cosmologists would agree that there is more to the cosmos than the enclave our telescopes can view. Observations indicate, in fact, that space is flat and infinite, like an endless plain. Therefore, if the number of possible combinations of particles is finite, anything that happens in our sector of space would occur again and again elsewhere in the infinite universe. It is like the adage that if you placed a million monkeys in front of a million keyboards and waited long enough, by sheer chance one of them would type out a passage from Shakespeare. Replace that with an infinite number of monkeys and an infinite number of keyboards; an infinite number of them would type out the same Shakespeare passage. Consequently we might think of infinite space as the home of an endless array of parallel realities. It is humbling to think that there may be myriad versions of each of us, living in parallel Earths, unaware of the others' lives.

Level II of Tegmark's scheme is based on eternal inflation. It constitutes the ever-growing set of bubble universes emerging from random quantum fluctuations. The existence of this level hinges on proving Linde's hypothesis that inflation leads to the endless production of bubble universes.

Level III consists of Everett's Many Worlds. The parallel universes in that level are in some ways the closest—alternative versions of our own reality. These would branch off each time a quantum process occurs. Each quantum transition would escort copies of each of us to various alternative realities. However, we could never travel "sideways" in time and visit the paths not taken. Therefore, Level III would be just as inaccessible as the other levels—perhaps even more so, given that there may be indirect ways of measuring the properties of regions outside the observable universe.

Tegmark reserved Level IV for his own speculative proposal: the set of all mathematical structures. It would contain an endless variety of universes with different mathematical laws. For example, in some of these the irrational number pi could have a different value, changing the very nature of geometry.

There is no physical reason to think that the laws of mathematics could differ in other spaces. On the contrary, many mathematical relationships are required for stable astronomical structures, such as planetary systems, to develop. The inverse-squared law of gravity, for example, ensures that the planets in the solar system don't spiral inward, toward the Sun, or fly off into deep space. If the principles of geometry changed in another universe, the law of gravity would likely be different, too, precluding, perhaps, the existence of worlds for beings to live on circling stars.

Nevertheless, it is conceivable that beings could survive with radically different bodies—perhaps with no bodies at all. Hypothetically, intelligence based on pure energy could thrive under a broader variety of astronomical arrangements. Therefore, although it is hard to envision universes with bizarre mathematical rules, conceivably they could exist and even harbor some form of life.

The Multiverse Family Reunion

In 2011, Susskind and Raphael Bousso offering an intriguing proposal designed to unite the multiverse theory of eternal inflation with the Many-Worlds Interpretation of quantum mechanics. In forging this union, they drew upon another alternative to the Copenhagen interpretation, called decoherence. Their goal was to offer a consistent picture of quantum transitions that excludes the idea of collapse while still replicating the probabilities and other results associated with that idea.

Decoherence, largely the work of physicists H. Dieter Zeh and Wojciek Zurek, is a deterministic alternative to quantum collapse that involves an irreversible interaction between a particle's quantum state and its environment. Environment in this context means everything in causal contact with that particle that could potentially influence it, including the observer taking the measurement. Before this transition the system is said to be in a pure state. For example, in the case of electron spin, the pure state would be a superposition of spin up and spin down. Once the system is observed, or otherwise interacts with its environment, the ensuing entanglement between the state and its surroundings would favor a transition into a new state that corresponds to a particular value of what is being measured. In the case of spin, that would be either up or down. We can think of this transition as a release of quantum information about the particle into the environment, whittling away all the options down to one, leading to what appears to be a classical (non-quantum) result. In other words, the state sheds all its ambiguity into the environment and becomes resolute. Although the transition wouldn't be instantaneous, as quantum collapse is thought to be, it would take place so quickly that the intermediate state couldn't be observed. Thus, even though it would take an exceedingly brief interval to complete, it would look like instant collapse.

Bousso and Susskind considered how decoherence theory would apply to the quantum state of the entire universe. If there was no external environment, decoherence couldn't occur, and the universe would remain in the same superposition of states forever. It would be like Schrödinger's cat, in a world without observers, persisting in its zombielike mixed state forever. In the case of a universe condemned to be eternally coherent (completely in superposition), every measurement taken of it would be hazy.

However, if one considers the universe to be part of a multiverse, such as in eternal inflation, decoherence could occur in a more clearly defined way. Any particular causally connected sector, which Bousso and Susskind called a "causal diamond," could reduce to a collapsed state by transferring information to other parts of the multiverse. Thus the multiverse would be the reservoir of all possible results of a quantum measurement. In essence, other bubble universes would play the role of Everett's Many Worlds. Bousso and Susskind call their theory "The Multiverse Interpretation of Quantum Mechanics."

The idea that the multiverse, with its completely inaccessible bubble universes, could be an essential ingredient of quantum physics is truly mind-boggling. We may not be able to visit alternative realities, but nevertheless they could still be influencing the world around us. If Bousso and Susskind's hypothesis is correct, our very lives could depend on quantum processes shaped by the existence of the multiverse.

When our universe ends someday, perhaps other sections of the multiverse would persist. As our own enclave of space nears its demise, perhaps our distant descendants could somehow transfer information about our society to a parallel universe, and preserve our civilization elsewhere. How much longer we have until the demise of the universe depends on whether its acceleration persists.

Will the universe end in a Big Rip, Big Stretch, Big Crunch, Big Bounce, or yet another apocalyptic scenario? Its fate depends on cosmological parameters astronomers are carefully trying to measure, such as precisely how its expansion and acceleration rates have changed over time. The ultimate destiny of everything around us—the myriad stars that speckle the sky and all the planets warmed by their nuclear furnaces—hangs in the balance.

14

How Will the Universe End?

With a Bang, Bounce, Crunch, Rip, Stretch, or Whimper?

I had a dream, which was not all a dream.
The bright sun was extinguish'd, and the stars
Did wander darkling in the eternal space,
Rayless, and pathless, and the icy earth
Swung blind and blackening in the moonless air . . .

The waves were dead; the tides were in their grave,
The moon their mistress had expir'd before;
The winds were withered in the stagnant air,
And the clouds perish'd; Darkness had no need
Of aid from them—She was the Universe.

—LORD BYRON, *"DARKNESS"*

The fate of the universe is a subject that hangs heavy on all of us. It is horrific to imagine all of our creations, from the paintings in the Louvre to the works of Shakespeare, crumbling into dust. If Earth were ever imperiled, at least we'd have the hope of looking elsewhere in space for habitable worlds. We could clamor to build high-speed spacecraft able to evacuate our planet and transport us to another world, where we could try to preserve our cultural treasures and traditions. We might even imagine a transgalactic civilization based on interstellar travel, bringing precious terrestrial knowledge to far-flung planets across the Milky Way. The end of the cosmos, however, would offer no such hope for continuation. The curtain would fall on human dreams and aspirations, leaving behind only an emptiness beyond emptiness.

"This is the way the world ends; not with a bang but a whimper," T. S. Eliot famously concluded his 1925 poem "The Hollow Men." The end of the universe could be catastrophic or drawn out, a booming crescendo of destruction or the soft murmurs of a protracted decline. Cosmologists talk not only about bangs and whimpers, but also about bounces, crunches, chills, freezes, and rips. Depending on the nature of the content of the universe, particularly its dark energy, cosmic demise could take many different forms.

Recent astronomical data, collected through WMAP and other sources, have sized up the value of omega: the ratio between the density of material in the universe and the critical density. Of the three possibilities for omega—greater than 1 for a closed universe, less than 1 for an open universe, and equal to 1 for a flat universe—all indications point to a value of 1.

Today the size of the observable universe is growing larger and larger. Conceivably, though, billions of years from now, the expansion of the universe could reverse course and its horizons could start shrinking. Doom would arrive in a Big Crunch that would squash the universe back down to a point.

Unless new data contradict the WMAP results, however, prospects for a Big Crunch seem highly unlikely. Theory suggests that a universe with an omega value of 1 would expand forever and never experience a Big Crunch. Nevertheless, a thorough examination of the fate of the universe requires including the Big Crunch along with other possible doomsday scenarios.

Rewinding the Big Bang

Philosophically, the Big Crunch scenario has always been a popular choice because of its satisfying symmetry. It would be somewhat like the Big Bang in reverse, leading back to a state of infinite density. Although, as physicists Raymond Laflamme and Don Page showed, time itself would not roll backward, space would switch its behavior from expansion to contraction. Eventually, everything the Big Bang built up, the Big Crunch would crush back down. A graph plotting how space develops during the final half of time would look like the mirror image of the initial half. The end of the universe would resemble the Big Bang in reverse, only with aged material rather than nascent energy and matter.

Psychologically, the Big Crunch offers the appeal of closure. In his play *As You Like It*, Shakespeare described the final stage of life as "second childishness and mere oblivion," conveying the view, however misguided, that old age is a kind of time reversal of childhood. Similarly, thinking that the universe will enter a "second childhood" provides a way of trying to close the circle and tie its chronicle together into a neat package.

Suppose, against today's odds, the Big Crunch scenario did transpire. At some point in the future, assuming our civilization still existed, astronomers would begin to notice that some galaxies beyond our Local Group were starting to slow their outward movements, as measured by decreasing redshifts. Eventually telescopes equipped with spectrometers would start to record Doppler blueshifts, meaning that galaxies were beginning to move toward us. Spectral lines of more and more galaxies would start to shift toward the blue end, signaling a growing stampede in our direction. Naturally, it wouldn't just be we who were the targets of the rampage—all points in space would be moving closer to one another. Eventually the blue-shifted radiation could prove energetic and intense enough to turn deadly and fry all forms of life. Finally, as space grew smaller and smaller, all planetary systems would be pulverized. From the time the contraction started until the ultimate crushing moments would be a solemn span of many billions of years.

The end state of the universe, in the event of a Big Crunch, would have aspects in common with black holes. In the final moments, the timelines of everything that exists in the universe would converge in a crushing singularity. Just as for a space traveler trapped within a Schwarzschild black hole's event horizon, there would be no escape.

As in the case of a large black hole, the Big Crunch would represent a high-entropy situation—only much more so, as it would collect the combined entropy of all the black holes and other material that existed at the end of time. The Big Bang, on the other hand, must have had vanishing (as close to 0 as possible) entropy. The discrepancy between the two offers a natural arrow of time, pointing from the low-entropy distant past to the high-entropy far future.

In 1979, Oxford cosmologist Roger Penrose suggested a way to represent the distinction between the two terminuses of time in general relativity. Einstein's equations establish a connection between the Einstein tensor, one way of representing the geometry of space-time, and the stress-energy tensor, delineating the matter and energy in a certain region or the whole universe. (A tensor is a mathematical entity that transforms in certain ways.) The equations depict how mass and energy warp the fabric of space-time. Penrose called upon a different mathematical entity, called the Weyl curvature tensor, as a means of pinning down the entropy of the universe. The Weyl tensor represents another way of describing the geometry of space-time that does not directly depend on the matter and energy distribution. It constitutes how tangled space-time is, rather than its warping. While warping depends directly on the amount of mass and energy in a region, twisting does not. Thus even if matter and energy abound at a particular moment in the history of the cosmos, the Weyl tensor could be 0.

In the Weyl Curvature Hypothesis (WCH), Penrose explained the low entropy at the start of time by linking it with the Weyl tensor and setting that to a low value. Specifically, he connected the Weyl tensor with the gravitational entropy—a measure of the disorder of gravity itself. Orderly gravity offers a regular, paradelike procession of the expansion of space; in other words, homogeneous growth from the Big Bang. That's essentially what astronomers have observed on the very largest scale, although the recent discoveries of dark flow, giant superclusters, and so forth have called large-scale homogeneity into question.

Penrose addressed the chaotic, high-entropy state of the end of the universe by showing that the components of the Weyl tensor would get larger and larger over time. Thus the growth of the Weyl tensor, equated with the increase of gravitational entropy, would offer a natural arrow of time's direction. The Weyl tensor would serve as the "clock" of the universe, recording how late and disordered it was in the history of the cosmos. The Big Crunch, if it transpired, would represent the sonorous final chimes of that grand timepiece.

Life in Deep Freeze

The universe evolves on its own terms, not to satisfy aesthetic symmetry arguments or psychological needs for closure. Based on observational evidence, crunch time for the universe doesn't seem programmed into its multibillion-year calendar. Rather, the ending seems more likely to be a prolonged deep freeze. Flat or open universes, expanding forever, also continue to cool. Such protracted, frigid declines have been called the Big Whimper, Big Freeze, or Big Chill. If, because of phantom energy, space ends up torn to shreds, then Big Rip is the term used. Teetering between the Big Whimper and Big Rip scenarios would be the case in which dark energy is well represented by the cosmological constant, meaning that space would spread out faster and faster, almost to the point of ripping—in what I call a Big Stretch.

As early as the mid-nineteenth century, Rudolf Clausius, formulator of the second law of thermodynamics, realized that ultimately the entire universe would run out of usable energy, a concept he called "heat death." Any mechanical process produces waste, so to keep it going it needs an external energy source such as coal, oil, solar energy, or another fuel source. Coal and oil derive from long-decayed plant material—from once-living vegetation that aeons ago collected energy from the Sun. Thus much of our usable energy (aside from nuclear and geothermal power) derives directly or indirectly from our parent star. After the Sun dies, we could attempt to derive energy from other sources, but ultimately these would run down, too. Eventually all stars are doomed to expend their nuclear fuel, turning into white dwarfs (and then burning out), neutron stars, or black holes. Gradually leaking energy by means of Hawking radiation, all black holes will eventually disintegrate. Thus, assuming the universe lasts long enough, everything is fated to become cold, lifeless, inanimate shards floating in an immense cosmic graveyard.

The temperature of space would continue to decline from its current frigid 2.73 degrees Kelvin (above absolute zero) to even closer to absolute zero. While absolute zero itself could never be reached, the cosmic temperature would approach it more and more. The ultimate stage of heat death would involve every physical system being in its lowest allowed quantum state.

Partly in reaction to a comment by physicist Steven Weinberg about the pointlessness of the universe, in 1979 unabashedly optimistic physicist Freeman Dyson decided to contemplate ways by which intelligence could survive despite the universal deep freeze. Dyson argued that intelligent living beings are instrumental to the future course of the universe, despite many scientists' desire to keep cosmology and life separate. As he wrote,

> It is impossible to calculate in detail the long-range future of the universe without including the effects of life and intelligence. It is impossible to calculate the capabilities of life and intelligence without touching, at least peripherally, philosophical questions. If we are to examine how intelligent life may be able to guide the physical development of the universe for its own purposes, we cannot altogether avoid considering what the values and purposes of intelligent life may be. But as soon as we mention the words value and purpose, we run into one of the most firmly entrenched taboos of twentieth-century science.[1]

Dyson argued that an open (or flat) universe would present more opportunities than a closed scenario. He found the closed possibility very bleak, as it would be almost impossible to prevent the frying and crushing ending. On the other hand, the open universe, in his opinion, would offer unlimited horizons and the possibility of indefinite survival of intelligent beings. He hypothesized that intelligent life would eventually shed its bodily form and transfer its consciousness to another medium, such as a computer or, more abstractly, an information storing and processing cloud of material. That collection of entities would learn how to slow its thinking and conserve its energy by greatly lowering its own temperature and hibernating as much as possible. He calculated that by spacing out their thoughts more and more, the intelligent beings of the future could expend only a finite amount of fuel in an infinite amount of time, thus persisting forever.

The situation would have some aspects in common with avant-garde composer John Cage's 1987 musical piece "As Slow as Possible." Cage died in 1992 without leaving specific instructions as to the longest amount of time in which it could be performed. Interpreters of the composition found a 639-year-old organ in Halberstadt, Germany, on which to play it. Taking the advice in its title to heart, because organs can be played extremely slowly, they decided to stretch the piece out to last 639 years. From September 2001 until February 2003, the bellows of the organ filled with air, and finally the first note was played. Each quarter note, instead of taking a few seconds, lasts two to four months. Changes in tone take place about once every year or two. Enthusiasts must hold their applause until the year 2640, when the performance is supposed to wrap up.

If Dyson is correct, future beings will follow the Halberstadt example and stretch out their creative processes in similar fashion. Like the spacing out of notes in Cage's piece, they will time their thoughts as slowly as possible, millions or even billions of years between each one. A long pause in a conversation won't be taken as offensive, but rather as a sign of economic energy use. In such a manner, humankind's symphony of creative thought could span the aeons.

In the late 1990s, University of Michigan researchers Fred Adams and Greg Laughlin elaborated on Dyson's scenario by contemplating how intelligence might survive during a prolonged cosmic decline. They divided the chronology of the universe into five distinct stages. The first, the "Primordial Era," was the age before star formation. The second, the "Stelliferous [star-filled] Era," represents the current period, in which stars shine through thermonuclear processes, and planets surround some of them. That era will draw to a close once most shining bodies expend their nuclear fuel and eject much of their outer material, while their cores implode to become white dwarfs, neutron stars, or black holes. Such relic objects will dominate the age that follows, dubbed by Adams and Laughlin the "Degenerate Era." ("Degenerate," in this case, is a technical term having to do with filled energy states; it does not reflect on the moral climate of the far future.) During that time, according to the researchers, protons could decay—a hypothesis supported by certain Grand Unified Theories linking the strong and electroweak interactions. Once all the protons decay, the "Black Hole Era" will commence in which no remnants will be left except for black holes. Finally, the black holes themselves will dissolve through the process of Hawking radiation, leading to the last stage of the universe, called the "Dark Era," when the universe is a dilute bath of ultracold elementary particles—particularly electrons, positrons, extremely-low-frequency photons, and decay products left over from earlier times. By then, according to Adams and Laughlin, the universe will be more than 10^{100} (1 followed by one hundred zeros) years old!

Following Dyson's idea of life slowing to accommodate the lower-energy resources of the far future, Adams and Laughlin paint an optimistic picture of the prospects for the long-term existence of intelligent beings. Recognizing that the current era has many favorable aspects for organic life forms (abundant stellar powerhouses such as the Sun), the authors caution nevertheless to be open-minded about the prospects for alternate living entities in other eras. For example, even in the Black Hole Era, perhaps living beings of a much different composition than carbon-based life could subsist in the extraordinarily frigid temperature of one ten-millionth of a degree Kelvin produced by Hawking radiation. These creatures would have extremely sluggish metabolisms fit for survival in the cosmic freezer—with thought processes operating billions of times slower than ours. Thus it is possible that our distant descendents will ring in the New Year of A.D. 1 tredecillion (10^{42}, or 1 followed by 42 zeros) by toasting each other with radiation cocktails meant to be sipped very, very slowly.

The Heart of a Lonely Universe

The 1998 discovery of cosmic acceleration came just on the heels of Adams and Laughlin's predictions, and proved to be a game-changer for prognostications of the far future. Unless acceleration someday abates, the Big Crunch and Big Whimper scenarios are out (at least in their purest forms), to be replaced with the frightening prospect of a Big Rip—or at least a Big Stretch. The major differences between the Big Rip and Big Whimper scenarios have to do with the prospects for maintaining communication with other parts of the universe, as well as the fate of space itself.

The term "Big Rip" was coined in a 2003 paper by Robert Caldwell—along with Caltech researchers Marc Kamionkowski and Nevin Weinberg—with the alarming title "Phantom Energy and Cosmic Doomsday." The authors describe the long-term fate of the universe subject to phantom energy—their term for a persistent form of dark energy with a w factor (ratio of pressure to density) that is less than negative 1. As they point out, such a strongly negative pressure contribution would become an increasingly large fraction of the substance of the universe. Space would expand faster and faster, pushing the galaxies away from one another like a leaf blower blasting away a pile of leaves. One by one, each remote galaxy will retreat past the cosmic horizon, never to be seen again. Eventually the only galaxies within telescopic reach will be members of the Local Group, such as Andromeda and the Magellanic Clouds.

As Lawrence Krauss of Arizona State University and Robert Scherrer of Vanderbilt University have pointed out, ironically, once all the galaxies except our Local Group have disappeared, it will appear as if the universe has "stopped expanding." That is because our current gauge of expansion is the behavior of more remote galaxies. If future life forms in the Milky Way lack historical accounts of the cluttered way the cosmos used to look, they might well conclude that our galaxy and its satellites form the bulk of the universe. It would be a throwback to the days before Edwin Hubble's discoveries, when astronomers thought the Milky Way was everything.

"In some sense it's poetic," said Krauss. "The future universe will look much like what people first thought it did when they first started thinking about cosmology."[2]

Phantom energy's influence would not cease, however, and it would start to break up the Local Group and even the Milky Way itself. More and more gravitational connections would be severed as the negative pressure substance occupied more and more of reality. Its supreme repulsion would overcome all of the attractive forces of nature, breaking everything up into its components. Eventually the garment of space itself would be torn apart like a moth-infested wool sweater. In shredding space's very fabric, the Big Rip would complete its rampage of destruction.

The timetable for such devastation depends on the value of the w factor. If it happens to equal –1.5, for example, it would take about

20 billion years from now for the Big Rip to transpire and space to be decimated. Some 60 million years before the end of the universe, the Milky Way would tear apart. The solar system would break up only three months before the end. Just 30 minutes before the Big Rip, our planet (Earth, if it still exists, or perhaps a successor world where our descendants have taken refuge) would explode. The end of atoms would precede the breakdown of space itself by only a tiny fraction of a second: 10^{-19} seconds, to be precise.

If the value of w is less negative, but still smaller than -1, then the phantom energy would take longer to complete its dastardly doings and the Big Rip would happen later. However, for w precisely equal to -1, matching the cosmological constant scenario, or w between 0 and -1, representing quintessence, there would likely never be a Big Rip. The finale of complete spatial destruction would be forever postponed.

Driving by a cosmological constant, or a form of matter with negative pressure that is not as strong as phantom energy, space would still expand faster and faster, in a Big Stretch. While it would never tear apart, it would still fling distant galaxies one by one over the cosmic horizon. In many billions of years, our Local Group of galaxies, including the Milky Way and Andromeda, would become isolated. Gradually, as entropy builds up and heat death sets in, our galaxy would age gracefully into a collection of stellar remnants such as white dwarfs, neutron stars, and black holes. Remote galaxies would be completely inaccessible and unobservable. The Big Stretch would combine the chilling end promised by the Big Whimper, with a hefty dose of loneliness for good measure. In space, no one could see us freeze!

The Thaw after the Frost

Whenever there are portents of cosmic winter there are hopes for cosmic spring. Could the buds of new life be buried in the snow of a deep freeze? The Steinhardt-Turok clash of the branes scenario (discussed in chapter 7) is one such mechanism for picturing how the universe could regenerate itself. By replacing the Big Bang with a Big Bounce, it connects the end of time with its beginning in an endless loop.

Roger Penrose has recently offered another type of cyclic scenario, firmly lodged in standard general relativity rather than in unseen extra dimensions. His model, called Conformal Cyclic Cosmology, combines the Weyl Curvature Hypothesis with the idea of conformal invariance. Conformal invariance occurs when certain physical or mathematical quantities are independent of scale. This is akin to a little boy having a cleft chin or another characteristic facial feature that remains for life. No matter how large he grows, it would always be there—the same shape, but just a different size. Similarly, there are certain symmetries and features in physics and math that depend on shape rather than size.

In 2003, Oxford mathematician K. Paul Tod, a former student of Penrose, found a way of expressing the WCH as a special starting condition for the Big Bang. The WCH states that the beginning of the universe has a minimum value of Weyl curvature. Weyl curvature represents gravitational disorder, meaning irregularities in the way space develops. If the Weyl curvature is 0, then the universe is purely isotropic.

In general relativity, matter and energy of the right type act as a kind of growth formula. If we think of the trails traced by points in the Big Bang expansion as strands of hair, the material in the universe makes them grow longer and longer. However, they could be straight or tangled. Weyl curvature governs how twisted they are, with 0 meaning completely unknotted, and a high value representing a tangled mess. Consequently, the WCH offers a kind of conditioner for the early universe that removes all the tangles and makes its initial growth completely isotropic, like perfectly straight hair.

Tod noted that pure isotropic geometries have conformal invariance. Even if their metrics ("maps" of the shortest distances between points) are multiplied by a scale factor, they represent the same shape. It is like a stylist looking at someone with long, straight hair, such as Marcia Brady of television's *The Brady Bunch*, and noting that it appears just as linear no matter how closely or far away he examines it.

Because of its clear way of representing isotropy, Tod offered conformal invariance at the start of the universe as a means of defining the WCH. That signifies that the first instant of the Big Bang looks the same on all scales, from the minute to the astronomical. What about the end of the universe? If it is full of black holes and other gravitationally complex objects, it would be quite tangled and have an overall high value of Weyl curvature. In that case it certainly wouldn't have conformal invariance, as a spot near a black hole would look different close up than from a distant vantage point that included other regions. In the end, size would matter as well as shape.

Penrose thought hard about the end of time and came to a startling revelation. Suppose his former collaborator Hawking was right and black holes radiated away. Given long enough time, none of those objects would be left in the end. What if protons and other elementary particles also decayed, as in Grand Unified Theories? There is no theory about electrons decaying, but it could be possible. What if, by a certain extremely late stage of the universe, nothing remained of mass? In that case the universe would be rapidly expanding but empty. Because, like a flat desert, it would look the same from all vantage points, it would have conformal invariance. Thus, except for its size, it would be identical to the starting point of time.

By splicing together the smooth beginning and end of time, like a vintage film loop, Penrose arrived at what he called an "outrageous new perspective."[3] The universe, he surmised, runs in a perfect cycle. The alpha is similar to the omega, the birth cry to the death knell. The only difference is the scale. But just as for a baby girl, her mother's face is her whole universe, it is possible that in the dawn and dusk of time, size has no place. The cosmos in its dying moments forgets its size—as it has no content, anyway—and is reincarnated as a tyke.

For the philosophical and psychological reasons mentioned earlier, such as the need for closure, cyclic models have much to offer. Penrose's scheme certainly ties together the extremes of time and size, presenting the history of the universe as a perfect package. However, as Penrose himself noted, it relies on the unproven notions that all massive elementary particles will eventually decay and that black holes will disintegrate and leave absolutely nothing behind. The latter idea has been the subject of much debate—stimulated by Hawking's suggestion that black holes spawn baby universes.

Nurseries for Baby Universes

Conservation laws are the security blankets of physics. They offer comfort that certain properties normally don't change. Total charge, mass/energy (the two are convertible), spin, and a number of other quantities remain constant under ordinary circumstances. However, extreme conditions, such as the singularities (points of infinite density) of black holes and the Big Bang, seem to wrest these blankets of constancy away.

From its inception, the Big Bang theory has implied a monumental violation of conservation laws at the fledgling instant of creation. After all, creation implies something from nothing. Cyclic models are generally attempts at continuity, offering transitions rather than sharp breaks. For example, the Steinhardt-Turok model derives its energy from interactions between branes. Penrose's cyclic cosmology, although breaking conversation of charge by allowing electrons to decay into pure energy, offers geometric continuity.

If the universe is not renewed, and meets a grisly end such as the Big Rip, or even a more quiescent demise, such as the Big Whimper, another type of continuity will be broken—our fate as a species. It is only natural to search for a way out. Aside from renewed cycles, might there be other avenues to permanence?

In 1988 Hawking examined the question of what happens to black hole information after such an object radiates away all of its energy. Does the black hole destroy the information, or could it somehow be preserved? The researchers developed a theory of baby universes that gestate in the gravitationally intense wombs of black holes and are born in another region of what is now called the multiverse. The "umbilical cords" of the offspring would be wormholes connecting them to our universe. Charge and other physical quantities could pass through these gateways, allowing black holes eventually to dissolve to nothingness.

Hawking imagined a spaceship traveling into a black hole and passing into a baby universe. From the point of view of the baby universe, it would have a new toy—a shiny rocket. However, the baby universe's timeline would be distinct from ours—a different kind of succession of moments that Hawking called "imaginary time." Sadly, the spaceship occupants' experience of time would involve swift passage toward a crushing death, so they would never perceive the entrance into the baby universe.

Thus, in the event that future intelligent beings wished to escape the fate of the cosmos, baby universes would not offer a safe escape route. On the other hand, they could offer a storage place for information we wished to preserve. It is conceivable, if they exist, that a civilization under threat could encode its chronicle in a way that could survive the journey.

Along with Laflamme, Hawking also put forth a more technical use for baby universes—to explain the low value of the cosmological constant. They calculated that the extra quantum uncertainty associated with their presence would drive the likely value of the cosmological constant way down, close to its observed quantity. Because there would be no way to investigate baby universes directly, this hypothesis is yet another multiverse idea (along with eternal inflation, many worlds, and so forth) that has remained speculative.

Several years after Hawking put forth his proposal, Lee Smolin cleverly noted a connection between baby universes and offspring in biology that suggested a cosmological process of natural selection. Smolin proposed that baby universes would have slightly different laws than their parent universes—similar to the Darwinian concept of variation. The children would grow up and create black holes and baby universes of their own. Then in a kind of survival of the fittest, the universes with physical laws that engender the most black holes would dominate over the others. These successful parents also would be the sort to produce ideal conditions for life, explaining why we are here.

Our universe, if it were cognizant, might be thinking of cute names for all its progeny. It might have already purchased the book *Beyond Astro and Celeste: Names for Cosmic Newborns*, and be leafing through all the possible appellations. However, a statement by Hawking at a June 2004 conference in Dublin may have put a damper on such plans. He conceded that he was wrong about black holes hiding their information by means of baby universes. Rather, the data escape with the radiation the black hole slowly releases. In changing his mind, he had lost a 1997 wager that he and Kip Thorne had jointly made about black hole information, and had to send a baseball encyclopedia to the winner of the bet, John Preskill of Caltech.

"There is no baby universe branching off as I once thought," Hawking said at the conference. "I'm sorry to disappoint science fiction fans but if information is preserved there is no possibility of using black holes to travel to other universes. If you jump into a black hole, your mass energy will be returned to our universe, but in a mangled form which contains the information about what you were like."

Alas, returning in a "mangled form" does not sound like a particularly promising way to escape the fate of the cosmos. If Hawking's reevaluation is correct, so much for a cosmological baby boom. On the other hand, maybe one universe is enough.

Reality's Last Stretch

It is intriguing to imagine escaping from this universe into other realms, and a future beyond the end of time. Limits are hard for probing minds to accept. Yet for any cosmological conjecture, it is wise to base its likelihood on observational evidence rather than on abstract speculations about unknown reaches.

The bulk of cosmological evidence points today toward the concordance picture of a universe that is flat and isotropic on its very largest scale and is a mixture of ordinary matter, cold dark matter, and dark energy. The dark energy seems to be a persistent contribution, resembling the cosmological constant term. Structure in the universe matches well the predictions of a very early epoch of inflation that turned minuscule quantum fluctuations into the clumps of material that eventually built up into stars and galaxies.

Putting these findings together offers a best-guess scenario about the future of space. Although it is comforting to imagine cycles of time, there is no direct evidence that anything like the Big Bang will ever transpire again. The Big Crunch, in which expansion switches gears and turns into contraction, would require a stunning turnaround in the role of dark energy. It would have to dissipate somehow or change its properties to allow acceleration to become deceleration. Furthermore, as space seems to be flat, meaning that the density of the universe is equal to the critical density, even if gravity were unhindered by dark energy it wouldn't have enough strength to crunch space back together.

The cyclic models of Steinhardt-Turok and Penrose each involve propositions yet to be proven. The Steinhardt-Turok model entails brane interactions in an extra dimension, or at least an energy field that simulates that effect. The Penrose model seems to require the decay of all massive particles, including electrons. Thus although a Big Bounce or other kind of cyclic model has philosophical advantages over the idea of creation from nothing, much experimental work needs to be done to justify its underpinnings. On the other hand, although inflation is widely accepted, theorists have yet to identify the inflaton (inflationary field) that caused it. They also cannot adequately explain why inflation isn't happening in myriad places and times, as in eternal inflation—or if it is ubiquitous, what explains our special conditions.

If we presume that the universe doesn't cycle, that leaves several possible ending scenarios. If dark energy fizzles out, as in a decaying form of quintessence, the Big Whimper could be in the cards. If dark energy is turbo-boosted, as in phantom energy, brace yourself (if you plan to be around for tens of billions of years) for the Big Rip! So far, evidence seems to point toward a middle course—continued acceleration of the cosmos that threatens to isolate our galaxy, but not enough to tear the fabric of space itself.

A Big Stretch demise of the universe, involving the domination of dark energy in its final age, would be less dramatic than many of the other possibilities. Civilization could survive for many billions of years, thriving on stellar radiation, and then perhaps many billions of years further, reducing its metabolism to a colder and colder state, as in Dyson's hypothesis. Following the poetic advice of Dylan Thomas, it might go feistily rather than gently into the "long, dark night."

Over the aeons, the Milky Way would occupy an increasingly central place in our scope because the other galaxies, as they receded, would be harder and harder to observe. Hence, if the human race ever wished to attempt intergalactic exploration, the time would be sooner rather than later. Ultimately, even if our culture couldn't hope to survive to future cycles or escape to baby universes, at least it could take pride in its grand achievements, including its mastery of the riddles of science. Until the last black holes evaporate, that would be our legacy.

15

What Are the Ultimate Limits of Our Knowledge about the Cosmos?

Come forth, O man; yon azure round survey,
And view those lamps which yield eternal day.
Bring forth thy glasses; clear thy wond'ring eyes:
Millions beyond the former millions rise:
Look farther;—millions more blaze from remoter skies.

—HENRY BAKER, "THE UNIVERSE" (1727)

Perched on a diminutive planet, orbiting an average star, situated in the periphery of a typical spiral galaxy, perhaps it is audacious of us to purport to understand the mechanisms of an immense, possibly infinite universe. Such aspirations seem even more absurd given that our own universe could well be the merest leaf in a multiverse of endless pages. But since no other sentient beings anywhere in the cosmos have stepped up to the task, as far as we know, someone has to do the job. So why not us?

Earth's diameter of about 7,900 miles represents but about .000000000000000002 percent of the diameter of the observable universe—estimated to be approximately 93 billion light-years. In other words, one would need to line up more than 40 million trillion Earths in a row to span the observable universe. We are but a mere speck in the celestial firmament, but we wish to know everything!

Ancient mariners traveled across the seas to explore uncharted lands. If we wished today to explore the universe that way, we would be stymied by the enormous distances involved. From our tiny enclave, mapping out the observable universe would seem a herculean task. Until science develops much more powerful means of propulsion, our chances of venturing out in spaceships and exploring a large portion of the galaxy, let alone rest of the observable universe, are minimal. Interstellar flight could well be centuries away. Before we are able to soar among the stars, we need to explore the cosmos from the comfort of home.

Fortunately, a torrent of light from parts of space up to billions of light-years away constantly rains down on Earth. In recent years, we have developed the tools to scan the outer reaches of space, collect the myriad pixels of luminous information, and voyage in our minds to the farthest places our telescopes can observe. Thanks to powerful astronomical instruments such as WMAP and the Hubble Space Telescope, astronomers have been blessed with increasingly precise, comprehensive, and meaningful data about the cosmos. To press the limits of knowledge even farther, a new generation of even higher-resolution telescopes has begun to take their place.

Changing of the Guard

On August 20, 2010, WMAP caught its final glimpse of the cosmic microwave background—the culmination of a magnificent nine-year run that has changed the history of science. Three weeks later it settled into its retirement home: a steady solar orbit. It is certainly not the last we'll hear from WMAP, however. Data collected by the satellite are still being analyzed and interpreted. A final report will likely be released in late 2012 or 2013.

The achievements of WMAP in pinning down cosmological parameters have been incomparable. Given the uncertainty in the age of the universe for most of scientific history, up until the late twentieth century, WMAP's relatively precise value of 13.75 billion years (plus or minus 100 million years or so) is a stunning triumph. Couple that incredible contribution with the surgical slicing of the contents of the universe into its major components—ordinary matter, dark matter, and dark energy—and it is clear that WMAP has revolutionized cosmology by making it a precise science.

While WMAP, and COBE before it, have offered ultrasensitive "ears" for "listening" to the background hiss, the Hubble Space Telescope has supplied extraordinarily sharp "eyes" for gazing into the depths of the knowable universe. The granddaddy of all space-based instruments used in cosmology, Hubble is still going strong after more than two decades of service. New equipment added in 2009, including the Wide Field Camera 3, have extended its life span even further. Among its crowning achievements is the discovery of dark energy, for which Perlmutter, Schmidt, and Riess were awarded the 2011 Nobel Prize. Hubble also has imaged seemingly empty patches of the sky, revealing the "deep field" to be speckled with faint galaxies; imaged numerous supernova bursts; and helped establish the age of the universe. More locally, it has served up new, detailed looks at the outer solar system, offered stunning pictures of nurseries for newborn stars, provided proof of the existence of black holes, helped identify planets circling distant stars, and delivered astronomical images for untold other purposes.

The relative newcomers on the scene have similarly unlocked exotic vaults of cosmic secrets. The Fermi Gamma-Ray Space Telescope, launched in 2008, has offered vital data about the gamma-ray sky, tracked gamma-ray bursts and gamma-ray emissions from active galaxies, and introduced a new mystery: the dragons of the gamma-ray fog. Are these cosmic dragons signs of dark matter interactions, evidence of cluster formation, or indications of another unknown phenomenon? Valiant astronomers continue in their quest to slay that beastly conundrum.

On May 14, 2009, the European Space Agency launched two new instruments from its spaceport in Kourou, French Guiana: the Planck satellite and the Herschel Space Observatory. The Planck mission has picked up where WMAP left off, in offering a precise mapping of the cosmic microwave background with unprecedented angular resolution. Planck carries a telescope with a 1.5-meter (approximately 5-foot) mirror. The mirror collects microwave radiation and channels it to two ultrasensitive detectors, called the Low Frequency Instrument (LFI) and High Frequency Instrument (HFI). These instruments, in tandem, cover a wide range of frequencies prominent in the CMB. While the LFI has twenty-two radio receivers tuned to four different-frequency channels, the HFI has fifty-two bolometric detectors that work by targeting the radiation toward a heat-sensitive material that changes its electrical resistance in a measurable way—acting as electric thermometers. Planck has performed splendidly so far, producing an incredibly detailed map of subtle temperature variations in the CMB.

Many important projects are hinging on the final Planck results—expected to be released in 2013. For example, the blotches in the cosmic background found in 2011 among the WMAP data by Peiris, Johnson, Feeney, and their collaborators, and suspected of being signs of primordial collisions between bubble universes, require higher-resolution data supplied by Planck. Once the Planck data are available, the team expects to render a speedy verdict on whether these patches are statistically significant. If so, then perhaps we will have met the universe's baby cousins, long since grown and moved away! With healthy skepticism about such far-reaching phenomena, the scientific community will likely reserve judgment until other groups corroborate the findings.

Planck also will be used to distinguish different inflationary models of the universe and investigate alternative ways of explaining its apparent large-scale uniformity, such as bouncing cosmologies. Higher-resolution CMB data will help astronomers map out more precisely how structure formed in the early universe. It will help us understand whether the universe is completely isotropic, looking the same in all directions, or has subtle anisotropies. In the latter case, Planck could offer clues as to the origins of such anisotropies. Finally, it will help nail down more exact values of the Hubble constant, the rate of acceleration, and other cosmological parameters.

Planck's launchmate Herschel is probing the far infrared region of the spectrum. With a 3.5-meter (more than 11-foot) primary mirror, Herschel is the largest space telescope launched to date. Its purpose is to map out some of the coldest objects in space, including the frigid interstellar medium. Undoubtedly it will offer important insights about the formation of galaxies and other aspects of the evolution of structure in the universe.

In applauding the space-based telescopes, we must not forget the numerous ground-based observatories around the world—from the Keck Observatory in Mauna Kea, Hawaii, to Gran Telescopio Canarias in La Palma, Canary Islands. Updated with precise digital cameras and adaptive optical systems designed to counter the effect of atmospheric distortion, these offer much sharper images than in years past. Taken in tandem, modern telescopes—scanning the skies in a vast range of visible and invisible frequencies—have inaugurated a true golden age of cosmology.

A Tangled Webb

In May 2009, the same week as the launch of the Planck and Herschel probes, NASA astronauts on board the space shuttle *Atlantis* completed the last anticipated service mission for the Hubble. Their valiant efforts helped prolong the life span of that grand instrument beyond even its record two decades. However, the shuttle program has ended—all but eliminating the possibility of further upgrades. Sadly, the gleaming telescope, source of so many stunning images and critical data from the far reaches of space, cannot keep going forever—especially without any new service missions planned.

Hubble's stunning results have whetted the scientific community's appetite for a deeper exploration of space. We'd like to view as close to the edge of the observable universe as possible. To understand how the first generation of stars developed in the close of the cosmic Dark Ages and how nascent galaxies came into being require pushing the very limits of telescopic observation. This calls for a larger, more powerful space telescope.

The James Webb Space Telescope, NASA's designated successor to the Hubble, has been in the planning stage since the 1990s. Originally called the Next Generation Space Telescope, it was renamed in 2002 after the second NASA administrator, James E. Webb, who died in 1992. Scheduled to be launched in 2018, the telescope will boast a 21-foot mirror and an enormous sunshield, designed to unfold once the telescope is safely in orbit some 1 million miles from Earth. It will survey the sky mainly in the infrared range, but also in the visual, scanning for important clues about the early development of galaxies and collecting information about planetary systems in formation. In short, it will offer a time machine to the infant years of the universe, when features such as stars and galaxies were in their initial stages.

Like many federal projects, its prospects have waxed and waned with the shifting tides of congressional budgetary decisions. To date, more than $3 billion has been invested in the program. The total projected cost has grown throughout the years to more than $8 billion. While such figures represent but a small portion of the federal budget, growing deficits have brought them to the attention of congressional cost-cutters.

On July 6, 2011, the House Appropriations Committee released the 2012 Commerce, Justice, Science Appropriations bill, delineating funding for NASA and other agencies. The bill recommended cutting funding for NASA's science programs and canceling the Webb Space Telescope. It cited mismanagement and cost overruns as the reasons for terminating the project.

The situation threatened to be a replay of the notorious cancellation of the Superconducting Super Collider (SSC), a planned particle accelerator that was axed by Congress in the 1990s before it was completed. Like the SSC, the Webb was under construction when the budgetary decision was made.

The Webb's giant primary mirror is composed of eighteen segments. These are made of beryllium, a substance known for its light weight, rigidity, and resistance to warping under extreme temperature changes. In June 2010, engineers began the process of coating the mirrors with thin layers of gold, ideal for reflecting infrared light gathered from the depths of space. The delicate procedure would take more than a year to complete. It would have been folly to cancel the Webb in the midst of constructing one of its principal components.

Fortunately, after months of negotiations, in November 2011 Congress voted for full funding of the Webb Space Telescope. Legislators recognized how vital the Webb is for future scientific discovery as well as for job creation (the telescope supports twelve hundred jobs). Supporters of the Webb Telescope are now optimistic that its 2018 launch target date will be met.

Where Is the Antimatter?

The Webb Space Telescope, Planck Satellite, and other astronomical instruments can take science only so far in probing the early
universe. To understand how the jumble of energy that was the Big
Bang stratified into familiar particles with a diverse range of properties requires linking astronomical discoveries with the findings of
particle physics. Particle physicists are attempting to discern the ultimate laws of nature that guided how symmetries shattered and led
to gaping differences in the subatomic world. The breaking of symmetries, in turn, influenced how the primitive cosmos developed.

_ One especially puzzling aspect of the Big Bang that scientists are
striving to comprehend is why the initial cosmic balance of matter and
antimatter broke down. If the universe began in a state of neutrality, the
ledger of positive and negative charge should have precisely canceled
out. Therefore, for each quantity of matter particles, such as electrons,
there should have been the same amount of oppositely charged antimatter counterparts, such as positrons. These would have annihilated
each other, creating a bath of pure radiation. Yet that's not what happened. Somehow the universe ended up with a vast excess of matter—
and precious little antimatter.

Could antimatter be hiding in distant stars and galaxies? To date, no
one has detected any large reserves of antimatter, let alone "antistars"
and "antigalaxies." Furthermore, nobody has found signs of the enormous amounts of energy that would be released if bodies of matter and
antimatter happened to collide. (We must distinguish antimatter from
dark matter, as while the former is well understood but scarce, the latter
is common but little understood.)

To see if any pockets of antimatter exist in space, the U.S. Department
of Energy, CERN, and an international team of researchers developed
a sensitive instrument called the Alpha Magnetic Spectrometer (AMS-
02). Currently docked to the International Space Station (ISS), it was
carried into space by the shuttle *Endeavour* in May 2011 on the second-
to-last mission before the shuttle program ended. It was the last voyage
for *Endeavour* itself before it was retired.

One reflection of the *Endeavour* voyage's importance to cosmology was the design of its mission patch. As NASA described it, "The

shape of the patch is inspired by the international atomic symbol, and represents the atom with orbiting electrons around the nucleus. The burst near the center refers to the Big Bang theory and the origin of the universe. The space shuttle *Endeavour* and space station fly together into the sunrise over the limb of Earth, representing the dawn of a new age, understanding the nature of the universe."[1]

The commander of the mission, Mark Kelly, faced an extraordinarily difficult decision. On January 8, 2011, only a few months before the launch was scheduled, Kelly's wife, Representative Gabrielle Giffords of Arizona, was shot in an attempted assassination. During her intense physical rehabilitation, Kelly needed to make the grim choice of whether to abandon the mission, or to resume training for the spaceflight while still helping her recover.

On February 4, Kelly announced his decision. "We have been preparing for more than eighteen months, and we will be ready to deliver the Alpha Magnetic Spectrometer to the International Space Station and complete the other objectives of the flight."[2]

Thanks to Kelly and his crew, the AMS-02 was carefully secured to the ISS, where it has begun collecting data. Within its innards is a sophisticated particle detector able to sniff out the presence of antihelium, if it exists, and measure its ratio to the amount of ordinary helium. Helium and antihelium nuclei would have the same mass but opposite charge: positive vs. negative. Because magnetic fields bend the paths of positive and negative charges in opposite directions, a powerful magnet within the detector is used to reveal the each nucleus. The AMS-02 is searching for evidence of dark matter as well as antimatter. Its particle-detecting prowess is such that it has been nicknamed the "LHC in space."[3]

Nobel laureate Sam Ting is the patient helmsman of the mission, having planned and headed it since the early 1990s. Even if it is unlikely that we'll find antimatter in space, he has argued that we need to know for sure whether it exists. As he remarked,

It might appear a long shot. However, there is no compelling reason why the Universe should be made of matter rather than an equal mix of matter and antimatter.[4]

A prototype called the AMS-01, launched in 1998 on board the space shuttle *Discovery*, capped the ratio of antihelium to helium in space to less than one part in a million. AMS-02 is one thousand times more precise than its predecessor, allowing it to gauge effectively whether antihelium exists in all of observable space. Time is running out for antimatter to reveal itself—if there is any out there at all.

Somehow, in the very early universe, a kind of revolution shattered the matter-antimatter balance. Matter trounced antimatter and became the undisputed king. Theorists have a good idea of what caused this upheaval. They conjecture that this breakdown of symmetry stems from processes mediated by the weak interaction that violate a condition called CP (Charge-Parity) symmetry. (Recall that we mentioned CP symmetry in chapter 8 in our discussion of axions.)

CP symmetry involves holding up particle decays to a set of twin mirrors. One of these is a left-right inversion called parity reversal. It is like taking a left-handed glove and replacing it with a right-handed glove. The other transformation, called charge conjugation, is changing all positive charges to negative charges, and vice versa. Effectively that means that charged particles are replaced with their antiparticle mates, and antiparticles with comparable particles. Theorists once thought that combining both operations and applying these to any physical situation, such as decays, would lead to an equivalent process that would happen with equal likelihood.

In 1964, researchers James Cronin and Val Fitch, along with their colleagues, made the extraordinary discovery of CP violation in certain weak decays. While observing the decay of a particle called the long-lived neutral kaon (also known as the K^0 meson), they recorded the by-products to be two oppositely charged pions (another meson type). CP conservation would predict only decays into three pions, not two. In honor of their groundbreaking finding, Cronin and Fitch received the 1980 Nobel Prize in Physics. Since the time of their work, numerous experiments have detected CP violation in weak decays (but never in the electromagnetic or strong interactions). As such CP-violating decays can build up an excess of particles over antiparticles, they are a natural mechanism to explain how such an imbalance arose in the early cosmos.

Within microseconds after the Big Bang, the universe was cool enough that photons no longer naturally transformed into particle-anti-particle duos. Existing particle-antiparticle pairs canceled each other and reverted to radiation. However, likely because of CP violation, an excess of particles remained. These included the quarks that would form protons and neutrons, along with electrons, neutrinos, and other particles. Matter, as we know it, could finally start to come together.

A vital project investigating the nature of CP violation is the Large Hadron Collider Beauty (LHCb) experiment. Conducted in one of the beam intersection points of CERN's Large Hadron Collider, the experiment measures the properties of decays involving particles and antiparticles possessing the b-quark ("beauty" or "bottom" quark) or b-antiquark. These help ascertain differences between matter and anti-matter, critical to understanding how the former came to dominate over the latter in the universe.

Another LHC project, called the A Large Ion Collider Experiment (ALICE) experiment, investigates a different question about matter in the early universe: its fiery, ultradense state called quark-gluon plasma. For normal matter at moderate temperatures, its elementary constitu-ents—protons, neutrons, and other types of particles—are composed of three quarks each (except for a type called mesons, which are quark-antiquark pairs). According to the theory of the strong force, quantum chromodynamics (QCD), these are bound together by gluons. The more quarks try to separate, the stronger the gluons yank them back, preventing them from ever breaking free. However, in the very early universe, when temperatures were a hundred thousand times hotter than at the core of the Sun, quarks are believed to have had enough energy to be free to move, forming fleeting combinations with gluons and other quarks in a quark-gluon plasma. At the LHC, part of the operating season is designated for collisions between lead ions, rather than the usual proton-proton collisions. This has already successfully yielded evidence of a quark-gluon plasma state. By studying the sizzling "quark soup" created for brief moments in lead ion collisions, scientists hope for a greater understanding of the state of the primordial cosmos.

Particle physics offers the prospects of supplying the missing pieces to the cosmological puzzle that astronomy is unable to provide. Though the Webb Telescope and other future astronomical tools will extend cosmology's reach to unprecedented depths of space and time, they won't be able see back in time to the initial moments of the Big Bang, well before the time when the universe became transparent to light. Therefore, only in tandem will particle physics and astronomy be able to resolve cosmology's deepest mysteries.

Into the Unknown

It is the best of times and the weirdest of times for cosmology. Ironically, although the field is in excellent shape in terms of instruments, data collection, and analysis, it must blushingly concede a lack of understanding of what makes up more than 95 percent of the cosmos. It is like a botanist with encyclopedic knowledge of 5 percent of all plants, gleaned from years of experience and meticulous measurements, who shrugs his shoulders when asked about the other types and says he has no idea.

The current standard model of the universe is a tripod with two of its three legs tenuously perched on loose soil. It rests on the existence of conventional matter, dark matter, and dark energy. Dark matter provides the necessary scaffolding for visible matter to assemble. Eventually, after all the galaxies have reached, more or less, their mature forms, dark energy triggers a notable speeding up of cosmic expansion. Of these three necessary supports, the latter two are wholly unknown entities—despite years of detection efforts—measured only indirectly through their actions on visible matter.

While we still don't know what composes dark matter and dark energy, there are numerous intriguing possibilities. Will WIMPs or axions reveal themselves in underground experiments, space-based tests, or the debris of particle colliders? Or will dark matter prove to be an illusion—simply a modification of the law of gravity?

Dark energy is even more elusive than dark matter. Still, the theories of holographic dark energy, quintessence, chameleon particles, and phantom energy offer intriguing possibilities to test. In tandem with the more-than-decade-old technique of plotting the energy profiles of supernovas, the Baryonic Acoustic Oscillations method being pioneered by the Sloan Digital Sky Survey and other groups offers a promising approach for mapping the impact of dark energy and eventually revealing its true nature.

One of the strangest new results is the presence of dark flow—the rush of clusters of galaxies toward a particular region of space. Could this represent evidence of a multiverse? Could the strange alignment known as the "Axis of Evil" or the great cold spots offer additional support for outside influences, or are they just statistical flukes, like Hawking's initials in the microwave sky? What of the blotches in the microwave background? Are they indications of primitive collisions between bubble universes? And what of the strange dragons in the gamma-ray fog?

If science somehow proves the existence of a multiverse, there is no shortage of theories to describe it. Are the other parts of the multiverse products of eternal inflation or another mechanism? Could there be connections between the multiverse and the Many Worlds Interpretation of quantum mechanics? Might wormholes lead to baby universes? What of extra dimensions? Will physics ever be able to establish whether our universe is a brane floating in a higher-dimensional bulk, colliding periodically with other branes? The boundary between science and speculation surely seems to be eroding as more and more far-reaching, difficult-to-test hypotheses are being put forth. If we couldn't actually visit the rest of the multiverse, how could we measure its properties with certainty? On the other hand, perhaps we need to press our imagination to its very limits to explain the bizarre new findings in cosmology.

How will the universe end? The latest findings seem to indicate that dark energy will play a dominant role in stretching out space and isolating galaxies from each other. Still, it appears that we have many billions of years left before the Milky Way becomes a hermit. Let's use that time well to probe as much of the cosmos as we can. Perhaps we will someday even contact intelligent counterparts from other sectors of space who are conducting their own sky searches.

We arrive at the close of our exploration of the observable universe and our brave push beyond its very frontiers into unknown realms. Now it is time to return to Earth and marvel at the beauty of our own planet and the remarkable scientific progress made by our own humble species. In probing the profound mysteries of the vast cosmos from such a tiny vantage point, perhaps the power of our minds offers the greatest wonder of all.

Acknowledgments

I would like to acknowledge the support of my family, friends, and colleagues for making this book possible. Thanks to the faculty and staff of University of Sciences in Philadelphia, including Russell DiGate, Suzanne Murphy, Elia Eschenazi, Bernard Brunner, Sergio Freire, Ping Cunliffe, Dorjderem Nyamjav, Tarlok Aurora, Laura Pontiggia, Carl Walasek, Babis Papachristou, Salar Alsardary, Ed Reimers, Lia Vas, Amy Kimchuk, Barbara Bendl, Jude Kuchinsky, Phyllis Blumberg, Kevin Murphy, Robert Boughner, Samuel Talcott, Alison Mostrom, Jim Cummings, Christine Flanagan, Kim Robson, Roy Robson, Justin Everett, Elizabeth Bressi-Stoppe, and Brian Kirschner for supporting and encouraging my research and writing.

Many thanks to Justin Khoury, Kate Land, and Will Percival for valuable insights about their respective research programs. I appreciate the support of the community of Philadelphia science writers, including Greg Lester, Faye Flam, and Mark Wolverton, and the encouraging words of writers, researchers, and other creative individuals with whom I have corresponded, including Michael Gross, Marcus Chown, Clare Dudman, Michael LaBossiere, Victoria Carpenter, Lisa Tenzin-Dolma, Cheryl Stringall, Joanne Manaster, and Jen Govey. I thank Linda Dalrymple Henderson, David Zitarelli, Thomas Bartlow, Paul S. Wesson, Roger Stuewer, David Cassidy, and Peter Pesic for their encouragement. Thanks also to my friends for their support and advice, including Michael Erlich, Fred Schuepfer, Pam Quick, Mitchell and Wendy Kaltz, Dubravko Klabucar, Simone Zelitch, Doug Buchholz, Kris Olson, Robert Clark, Elana Lubit, Carolyn Brodbeck, Marlon Fuentes, Kumar Shwetketu Virbhadra, Steve Rodrigue, Mark Singer, and Robert Jantzen.

I appreciate the useful suggestions and advice of the staff at John Wiley & Sons, including my editors, Eric Nelson and Connie Santisteban, their editorial assistant, Rebecca Yeager, and my production editor, Richard DeLorenzo. I am grateful for the vital help and support of my agent, Giles Anderson.

Above all, thanks to my family for their unwavering love and support, including my wife, Felicia, who has been a source of great advice; my sons, Eli and Aden; my parents, Stan and Bunny; my in-laws, Joe and Arlene; along with the Antners, the Kesslers, and the Batoffs, Shara Evans, Lane and Jill Hurewitz, Richard, Anita, Emily, Jake, Alan, Beth, Tessa, and Ken Halpern, and Aaron Stanbro.

Notes

1. How Far Out Can We See?

1. J. Richard Gott III et al., "A Map of the Universe," *Astrophysical Journal* 624 (2005): 463.
2. Charles Misner, Kip Thorne, and John Wheeler, *Gravitation* (New York: W. H. Freeman, 1973), p. 5.

2. How Was the Universe Born?

1. Ralph Alpher, Hans Bethe, and George Gamow, "The Origin of Chemical Elements," *Physical Review* 73 (1948): 803–804.
2. E. Margaret Burbidge, Geoffrey Burbidge, William Fowler, and Fred Hoyle, "Synthesis of the Elements in Stars," *Reviews in Modern Physics* 29 (1957): 547–650.
3. Robert H. Dicke, P. James E. Peebles, Peter G. Roll, and David T. Wilkinson, "Cosmic Black-Body Radiation," *Astrophysical Journal* 142 (1965): 414–419.
4. Arno A. Penzias and Robert W. Wilson, "A Measurement of Excess Antenna Temperature at 4080 Mc/s," *Astrophysical Journal* 142 (1965): 419–421.
5. J. Richard Gott III et al., "A Map of the Universe," *Astrophysical Journal* 624 (2005): 463.
6. John C. Mather, Lyman Page, and P. James E. Peebles, "David Todd Wilkinson," *Physics Today* 56 (May 2003): 76.

3. How Far Away Will the Edge Get?

1. Adam Riess, "Logbook: Dark Energy," *Symmetry Magazine* 4 (October/ November 2007): 37.
2. Michael Turner, "Explained in 60 Seconds," *CAP Journal* 2, no. 2 (February 2008): 8.
3. Kate Land, correspondence with the author, September 14, 2010.
4. Timothy Clifton, Pedro Ferreira, and Kate Land, "Living in a Void: Testing the Copernican Principle with Distant Supernovae," *Physical Review Letters* 101 (2008): 131302.
5. Ibid.
6. Adam Riess, quoted in "NASA's Hubble Rules Out One Alternative to Dark Energy," news release, Space Telescope Science Institute, March 14, 2011, http://hubblesite.org/newscenter/archive/releases/2011/08. Last accessed April 9, 2012.
7. Alex Filippenko, quoted in Robert Sanders, "New Hubble Treasury to Survey First Third of Cosmic Time, Study Dark Energy," UC Berkeley press release, March 15, 2010, http://www.berkeley.edu/news/media/ releases/2010/03/15_hubble_treasury.shtml. Last accessed April 9, 2012.
8. J. Richard Gott III et al., "A Map of the Universe," *Astrophysical Journal* 624 (2005): 463.

4. Why Does the Universe Seem So Smooth?

1. Alan Guth, "Inflationary Universe: A Possible Solution to the Horizon and Flatness Problems," *Physical Review D* 23 (1981): 347.
2. Malvina Reynolds, *Little Boxes and Other Handmade Songs* (New York: Oak Publications, 1964).
3. Alan Guth, "Inflation," in *Carnegie Observatories Astrophysics Series*, vol. 2, *Measuring and Modeling the Universe*, ed. W. L. Freedman (Cambridge, UK: Cambridge University Press, 2004), p. 49.

5. What Is Dark Energy?

1. Lee Smolin and George F. R. Ellis, "The Weak Anthropic Principle and the Landscape of String Theory," preprint, January 2009, http://arxiv .org/abs/0901.2414.
2. George F. R. Ellis, Ulrich Kirchner, and W. R. Stoeger, "Multiverses and Physical Cosmology," *Monthly Notices of the Royal Astronomical Society* 347 (2004): 921–936.
3. Paul Steinhardt, interview with the author, Princeton University, November 5, 2002.

4. Paul Steinhardt, "The Quintessential Universe," Texas Symposium on Relativistic Astrophysics, Austin, December 2000. Reported in Christopher Wanjek, "Quintessence: Accelerating the Universe," *Astronomy Today*, http://www.astronomytoday.com/cosmology/quintessence.html.

5. Justin Khoury, correspondence with the author, September 13, 2011.

6. Ibid.

7. Will Percival, correspondence with the author, September 3, 2010.

6. Do We Live in a Hologram?

1. John Archibald Wheeler with Kenneth Ford, *Geons, Black Holes, and Quantum Foam* (New York: W.W. Norton & Company, 1998), p. 298.

2. Craig Hogan, "Holographic Noise in Interferometers," Purdue University Colloquium, March 2010.

3. Max Tegmark, Angelica de Oliveira-Costa, and Andrew Hamilton, "A High Resolution Foreground Cleaned CMB Map from WMAP," *Physical Review D* 68 (2003): 123523.

7. Are There Alternatives to Inflation?

1. Paul Steinhardt, interview with the author, Princeton University, November 5, 2002.

2. Linda Dalrymple Henderson, *The Fourth Dimension and Non-Euclidean Geometry in Modern Art* (Princeton: Princeton University Press, 1983).

3. Petr Hořava and Edward Witten, "Heterotic and Type I String Dynamics from Eleven Dimensions," *Nuclear Physics B* 460 (1996): 506.

4. Justin Khoury, correspondence with the author, September 13, 2011.

5. Paul Steinhardt and Neil Turok, talk at Dark Matter 2004, Santa Monica, California, February 18–20, 2004.

6. Paul Steinhardt, talk at New Horizons in Particle Cosmology: The Inaugural Workshop of the Center for Particle Cosmology, University of Pennsylvania, December 11, 2009.

8. What Builds Structure in the Universe?

1. C. E. Aalseth et al., "Results from a Search for Light-Mass Dark Matter with a P-type Point Contact Germanium Detector," *Physical Review Letters*, submitted February 2010, http://arxiv.org/abs/1002.4703.

2. Juan Collar, "Juan Collar on Dark Matter Detection," *Cosmic Variance: Discover Magazine*, April 21, 2008, http://blogs.discovermagazine.com/cosmicvariance/2008/04/21/guest-post-juan-collar-on-dark-matter-detection/. Last accessed April 9, 2012.

9. What Is Tugging on Galaxies?

1. Mike Hudson, "Research Profile," University of Waterloo website, http://science.uwaterloo.ca/research/profiles/mike-hudson. Last accessed October 23, 2011.

2. A. Kashlinsky et al., "A Measurement of Large-Scale Peculiar Velocities of Clusters of Galaxies: Results and Cosmological Implications," *Astrophysical Journal Letters* 686, no. 2 (October 2008): L49.

3. Carl Sagan, *Cosmos,* 1980 television series.

10. What Is the "Axis of Evil"?

1. Stephen Hawking, interviewed in the film *A Brief History of Time* (1991).

2. C. L. Bennett et al., "Seven Year Wilkinson Microwave Anisotropy Probe (WMAP) Observations: Are There Cosmic Microwave Background Anomalies?," *Astrophysical Journal Supplement Series,* submitted January 2010, http://arxiv.org/abs/1001.4758.

3. Ibid.

4. Kate Land in Karen Masters, "She's an Astronomer: Kate Land," Galaxy Zoo blog, September 1, 2009, http://blogs.zooniverse.org/galaxy-zoo/2009/shes-an-astronomer-kate-land/. Last accessed April 9, 2012.

5. George Ellis, "Note on Varying Speed of Light Cosmologies," *General Relativity and Gravitation* 39, no. 4 (April 2007): 511–520.

6. Max Tegmark, Angelica de Oliveira-Costa, and Andrew Hamilton, "A High Resolution Foreground Cleaned CMB Map from WMAP," *Physical Review D* 68 (2003): 123523.

7. Kate Land, correspondence with the author, September 14, 2010.

8. Ibid.

9. Ibid.

10. Ibid.

11. L. J. Hall and Y. Nomura, "Evidence for the Multiverse in the Standard Model and Beyond," *Physical Review D* 78 (2008): 035001.

12. Hiryana Peiris, interviewed by Jason Palmer, "Multiverse Theory Suggested by Microwave Background," BBC News, August 3, 2011.

11. What Are the Immense Blasts of Energy from the Farthest Reaches of Space?

1. Sean Farrell, "Extreme X-ray Source Supports New Class of Black Hole," University of Leiceister press release, September 8, 2010.

2. Keith Cowing, "Compton Gamma Ray Observatory Crashes on Earth," *SpaceRef*, June 4, 2000, http://www.spaceref.com/news/viewnews .html?id = 153. Last accessed April 9, 2012.

3. Stan Woosley quoted in Louise Donahue, "Scientists Part of Team Decoding Gamma-Ray Burst Mystery," UC Santa Cruz press release, June 18, 2003, http://news.ucsc.edu/2003/06/366.html. Last accessed April 9, 2012.

12. Can We Journey to Parallel Universes?

1. Stephen W. Hawking, "Chronology Protection Conjecture," *Physical Review D* 46 (1992): 603.

2. R. Paul Butler, quoted in Seth Borenstein, "Could 'Goldilocks' Planet Be Just right for life?," Associated Press, September 29, 2010.

3. Alejandro Jenkins, quoted in Anne Trafton, "Life beyond Our Universe," *MIT News*, February 22, 2010.

4. Robert Jaffe, quoted in Anne Trafton, "Life beyond Our Universe," *MIT News*, February 22, 2010.

13. Is the Universe Constantly Splitting into Multiple Realities?

1. Walter J. Moore, *Schrödinger: Life and Thought* (New York: Cambridge University Press, 1989), p. 233.

2. Ibid., p. 294.

3. Erwin Schrödinger, "The Present Situation in Quantum Mechanics: A Translation of Schrödinger's 'Cat Paradox Paper,'" *Proceedings of the American Philosophical Society* 124 (1980):323–338.

4. Eugene Shikhovtsev, "Biographical Sketch of Hugh Everett III," http://space.mit.edu/home/tegmark/everett/everettbio.pdf. Last accessed April 9, 2012.

14. How Will the Universe End?

1. Freeman Dyson, "Time without End: Physics and Biology in an Open Universe," *Reviews of Modern Physics* 51, no. 3 (July 1979): 448.

2. Lawrence Krauss quoted in Anne Minard, "Future Universe Will 'Stop Expanding,' Experts Suggest," *National Geographic News*, June 4, 2007, http://news.nationalgeographic.com/news/2007/06/070604-universe .html. Last accessed April 9, 2012.

3. Roger Penrose, "Before the Big Bang: An Outrageous New Perspective and Its Implications for Particle Physics," *Proceedings of 10th European*

Particle Accelerator Conference (EPAC 06), Edinburgh, Scotland (June 26–30, 2006): 2759–2767.

15. What Are the Ultimate Limits of Our Knowledge about the Cosmos?

1. "STC-134 Mission Patch," NASA, http://www.nasa.gov/mission_pages/shuttle/shuttlemissions/sts134/multimedia/gallery/134patch_prt.htm. Last accessed April 9, 2012.

2. Michael Curie and Nicole Cloutier-Lemasters, "NASA Astronaut Mark Kelly Resumes Training for STS-134 Mission," NASA press release 11–036, http://www.nasa.gov/home/hqnews/2011/feb/HQ_11–036_Kelly_Returns.html. Last accessed April 9, 2012.

3. Jonathan Amos, "How the 'LHC in Space' Lost Its British 'Engine,'" BBC News, August 27, 2010, www.bbc.co.uk/blogs/thereporters/jonathanamos/2010/08/how-the-grand-space-experiment.shtml. Last accessed April 9, 2012.

4. Sam Ting quoted in Marcus Chown, "Worlds beyond Matter," *New Scientist* (August 3, 1996): 36.

Further Reading

Technical works are marked with an asterisk.

Adams, Fred, and Greg Laughlin. *The Five Ages of the Universe: Inside the Physics of Eternity.* New York: Free Press, 1999.

Barrow, John. *The Constants of Nature: From Alpha to Omega—the Numbers That Encode the Deepest Secrets of the Universe.* New York: Pantheon, 2003.

Bartusiak, Marcia. *The Day We Found the Universe.* New York: Pantheon, 2009.

———. *Through a Universe Darkly: A Cosmic Tale of Ancient Ethers, Dark Matter, and the Fate of the Universe.* New York: HarperCollins, 1993.

Carroll, Sean. *From Eternity to Here: The Quest for the Ultimate Theory of Time.* New York: Dutton, 2010.

Chown, Marcus. *The Afterglow of Creation.* Herdon, VA: University Science Books, 1996.

———. *The Matchbox That Ate a Forty-Ton Truck: What Everyday Things Tell Us about the Universe.* London: Faber & Faber, 2010.

———. "Our World May Be a Giant Hologram." *New Scientist* 2691 (January 15, 2009).

———. *The Universe Next Door: The Making of Tomorrow's Science.* New York: Oxford University Press, 2003.

Croswell, Ken. *The Universe at Midnight: Observations Illuminating the Cosmos.* New York: Free Press, 2001.

Davies, Paul, and John Gribbin. *The Matter Myth: Dramatic Discoveries That Challenge Our Understanding of Physical Reality.* New York: Simon & Schuster, 1992.

Ferris, Timothy. *The Whole Shebang: A State-of-the-Universe(s) Report.* New York: Simon & Schuster, 1997.

Gates, Evalyn. *Einstein's Telescope: The Hunt for Dark Matter and Dark Energy in the Universe.* New York: W. W. Norton, 2009.

Goldberg, Dave, and Jeff Blomquist. *A User's Guide to the Universe: Surviving the Perils of Black Holes, Time Paradoxes, and Quantum Uncertainty.* Hoboken, NJ: John Wiley & Sons, 2010.

Goldsmith, Donald. *The Runaway Universe: The Race to Discover the Future of the Cosmos.* Reading, MA: Perseus, 2000.

Gott, J. Richard, and Robert J.Vanderbei. *Sizing Up the Universe: The Cosmos in Perspective.* Washington: National Geographic, 2010.

Greene, Brian. *The Elegant Universe: Superstrings, Hidden Dimensions, and the Quest for the Ultimate Theory.* New York: Vintage, 2000.

———. *Fabric of the Cosmos: Space, Time, and the Texture of Reality.* New York: Alfred A. Knopf, 2004.

———. *The Hidden Reality: Parallel Universes and the Deep Laws of the Cosmos.* New York: Alfred A. Knopf, 2011.

Gribbin, John. *In Search of the Multiverse: Parallel Worlds, Hidden Dimensions, and the Ultimate Quest for the Frontiers of Reality.* Hoboken, NJ: John Wiley & Sons, 2010.

Guth, Alan. *The Inflationary Universe: The Quest for a New Theory of Cosmic Origins.* Reading, MA: Perseus, 1998.

Halpern, Paul. *Collider: The Search for the World's Smallest Particles.* Hoboken, NJ: John Wiley & Sons, 2009.

———. *Cosmic Wormholes: The Search for Interstellar Shortcuts.* New York: E. P. Dutton, 1992.

———. *The Cyclical Serpent: Prospects for an Ever-Repeating Universe.* New York: Plenum, 1995.

———. *The Great Beyond: Higher Dimensions, Parallel Universes and the Extraordinary Search for a Theory of Everything.* Hoboken, NJ: John Wiley & Sons, 2004.

———. *Structure of the Universe.* New York: Henry Holt, 1996.

Halpern, Paul, and Paul Wesson. *Brave New Universe: Illuminating the Darkest Secrets of the Cosmos.* Washington, DC: National Academies Press, 2006.

Harrison, Edward. *Cosmology: The Science of the Universe.* New York: Cambridge University Press, 2000.

Hawking, Stephen. *Black Holes and Baby Universes and Other Essays.* New York: Bantam, 1993.

———. *A Brief History of Time.* New York: Bantam, 1987.

———. *The Universe in a Nutshell.* New York: Bantam, 2001.

Hawking, Stephen, and Leonard Mlodinow. *The Grand Design*. New York: Bantam, 2010.

Kaku, Michio. *Hyperspace: A Scientific Odyssey through Parallel Universes, Time Warps, and the Tenth Dimension*. New York: Oxford University Press, 1994.

———. *Parallel Worlds: A Journey through Creation, Higher Dimensions, and the Future of the Cosmos*. New York: Doubleday, 2004.

Kirshner, Robert. *The Extravagant Universe: Exploding Stars, Dark Energy, and the Accelerating Cosmos*. Princeton, NJ: Princeton University Press, 2004.

*Kolb, Edward, and Michael Turner. *The Early Universe*. Reading, MA: Addison-Wesley, 1990.

*Land, Kate, and João Magueijo. "The Axis of Evil." *Physical Review Letters* 95 (2005): 071301.

*Miralda-Escudé, Jordi. "The Dark Age of the Universe." *Science* 300, no. 5627 (2003):1904–1909.

Peebles, P. James E., Lyman A. Page, and R. Bruce Partridge, eds. *Finding the Big Bang*. Cambridge, UK: Cambridge University Press, 2009.

Penrose, Roger. *Cycles of Time: An Extraordinary New View of the Universe*. London: Bodley Head, 2010.

———. *The Road to Reality: A Complete Guide to the Laws of the Universe*. New York: Alfred A. Knopf, 2005.

Randall, Lisa. *Warped Passages: Unraveling the Mysteries of the Universe's Hidden Dimensions*. New York: HarperCollins, 2005.

*Rowan-Robinson, Michael. *Cosmology*. New York: Oxford University Press, 1996.

Sagan, Carl. *Cosmos*. New York: Bantam, 1985.

Singh, Simon. *Big Bang: The Origin of the Universe*. New York: Fourth Estate, 2004.

Smolin, Lee. *The Life of the Cosmos*. New York: Oxford University Press, 1999.

———. *The Trouble with Physics: The Rise of String Theory, the Fall of a Science, and What Comes Next*. New York: Houghton Mifflin Harcourt, 2006.

Steinhardt, Paul, and Neil Turok. *Endless Universe: Beyond the Big Bang*. New York: Doubleday, 2007.

Susskind, Leonard. *The Cosmic Landscape: String Theory and the Illusion of Intelligent Design*. New York: Back Bay Books, 2006.

Weinberg, Steven. *The First Three Minutes: A Modern View of the Origin of the Universe*. New York: Basic, 1993.

White, Michael, and John Gribbin. *Stephen Hawking: A Life in Science*. New York: E. P. Dutton, 1992.

Yau, Shing-Tung, and Steve Nadis. *The Search for Inner Space: String Theory and the Geometry of the Universe's Hidden Dimensions*. New York: Basic, 2010.

Index